ROADLESS

Now lads, I venture to tell you that I think you will live to see the day when land-cars will supersede almost all methods of conveyance in roadless countries - when armies will be moved across country and roadless traction will become the chief means of commercial movement for all undeveloped land.

Major General J. F. C. Fuller

Observe how system into system runs,
What other planets circle other suns *Alexander Pope*

ROADLESS

STUART GIBBARD

FARMING PRESS

First published 1996

Copyright © Stuart Gibbard 1996

All rights reserved. No parts of this publication may be reproduced,
stored in a retrieval system, or transmitted, in any form or by any means,
electronic, mechanical, photocopying, recording or otherwise,
without prior permission of Farming Press.

ISBN 0 85236 344 3

A catalogue record for this book is available from the British Library

**Published by Farming Press
Miller Freeman Professional Ltd
2 Wharfedale Road, Ipswich IP1 4LG, United Kingdom**

Distributed in North America
by Diamond Farm Enterprises, Box 537, Alexandria Bay, NY 13607, USA

Front cover photograph: Roadless 115
Back cover photographs: top, Super-Sentinel Roadless Steam Tractor;
left, Case C on Roadless DG4 half-tracks; right, Roadless 9804
Frontispiece: Roadless 115
Picture credits: Quadrant Picture Library p44 (top), p48 (middle)

Cover design and book layout by Liz Whatling
Printed and bound in Great Britain by Jarrold Book Printing, Thetford, U.K.

Contents

Acknowledgements … 6

Preface … 7

Chapter 1. From Swords to Plough Shares – 1919 to 1930 … 8

Chapter 2. Tractors on Tracks – 1925 to 1938 … 38

Chapter 3. War and Peace – 1939 to 1957 … 58

Chapter 4. Four-wheel Drive Developments – 1956 to 1964 … 88

Chapter 5. Power on the Land – 1965 to 1975 … 108

Chapter 6. Change and Decline – 1975 to 1983 … 138

Epilogue – 1983 to 1996 … 168

Appendices

  1 Selene and Four-wheel Traction … 172

  2 J.J. Thomas Tractors … 178

  3 Roadless Applications … 180

Useful Addresses … 185

Bibliography … 186

Index … 187

Acknowledgements

This book could never have been written without the assistance of the Men of Roadless, notably Ven Dodge, Vic Crockford, Mervyn Ford, Roger Haynes and Arthur Battelle – I owe them all my sincere gratitude.

They have all responded to my need for information with a thoroughness and enthusiasm that reflects the dedication and true strength behind the Roadless company. These people wanted the Roadless story telling as much as I wanted to write it. Between them, they have allowed me access to unique personal, private and company records, documentation and photographs which have given me an unprecedented insight into the workings of Roadless Traction.

Furthermore, they have given me much of their valuable time and shared with me their memories and extensive knowledge of the company and its products. They have all been available at the end of the telephone to patiently answer my enquiries, and I have always been made very welcome when visiting to talk (or, more often, listen) about Roadless, and as mornings extended into late afternoons, I often received both food for thought and food to eat. I remember one session with Ven Dodge to look at a few (!) photographs lasted seven hours without a break. Vic Crockford and Ven Dodge have also kindly read my manuscripts and put me back on the rails where necessary.

To Ven Dodge, I owe special thanks. Not only was he the inspiration behind this book, he has dug deep to find out whatever information he could for me with the same methodical and precise manner with which he ran the sales side of Roadless. Ven was also the company photographer, and he has used his considerable expertise in this field to provide me with first-class prints and copies from old, and often deteriorating, glass plates, negatives and photographs for use in the book. I am sure that the finished results bear testimony to his skill.

I am especially grateful to David Fletcher, the librarian for the Tank Museum at Bovington in Dorset, for help with much of the early information on Philip Johnson's tank developments, and also for kindly allowing me permission to use several tank and military vehicle photographs from the Bovington collection. If all museums responded as promptly and efficiently as the Tank Museum, research would be so much easier. It has to be said though that David Fletcher is also a devotee of Lieutenant-Colonel Johnson's work.

I would also like to thank Graham Melvin for the Jewells side of the story, and William Fuller for detailing the history of Four Wheel Traction. For information on Rover, I was kindly assisted by James Taylor and Graham Robson. Once again I am indebted to John Blackbeard for the loan of several photographs from *Arable Farming*, and also to David Cousins, features editor of *Farmers Weekly* (Reed Business Publishing) for permission to use a cutaway drawing from *Power Farming*. As usual, I am grateful to my wife, Sue – this time for managing to hoover round several cardboard boxes of Roadless records without disturbing me unduly.

I would also like to thank the following people who have lent photographs, provided information, or helped in some way – Peter Anderson, James Baldwin, David Bate, Colin Bennett, Bob Cavill, Robert Fearnley, David Hutchinson, Bruce Keech, Robin Ketley, Bill King, Ian Liddell, Peter Love, Tom Lowther, Richard Martin, Andrew Offer, Eric Sixsmith, Geoff Warren, David Woods.

Preface

To many the name Roadless is synonymous with four–wheel drive tractors. Yet four-wheel drive production is only the tip of the iceberg - while this was a very important part of the company's business, it occupied not quite thirty years of Roadless's sixty-four year history. For nearly forty years the company concentrated on building tracked vehicles, but even this is only part of the story and does not account for the many specialised avenues of design and innovation that the Roadless entered into. Several turned out to be cul-de-sacs, but others led on to important product developments that were unique for such a small company.

I originally embarked on the Roadless story with the intention of including it as a chapter in my previous book, *Ford Tractor Conversions*. It soon became clear that, not only would a complete history of Roadless never fit in the confines of one chapter, but to write about it under this heading was misleading. The bulk of Roadless's tractor conversions were indeed based on Ford skid-units, but very many other makes of machine were also converted at Hounslow and Sawbridgeworth. Furthermore, the company built much more than just tractors – including many products and developments so specialised that general categorisation is impossible.

Roadless was only a tiny company, and yet it was continually searching for new ideas and was at the forefront of design and innovation in many of the fields of transport and agricultural engineering; it pioneered the development of the half-track, and probably produced more crawler tractors than any other British company before the Second World War. Four-wheel drive tractors were coming out of Hounslow before many other manufacturers had even recognised the importance of the system. Even during the later years, the company's drive and creativity was undiminished, and in 1979 it brought out the first British, and probably the world's first, hydrostatic forestry tractor.

The history of Roadless is an important story that has never been fully told. This book is the result of my humble attempt to try and put that history into perspective. I make no excuses for including details of much of the early life and work of Philip Johnson. Lieutenant-Colonel Johnson was the founder of Roadless Traction, and was the driving force behind its first forty-odd years of existence. In the Philip Johnson Memorial Lecture to the Institute of Agricultural Engineers in November 1966, James A. Cuthberston referred to Johnson as 'a Pioneer and Natural Gentleman'. This brings to mind the words 'Inventor and Pioneer' used as the title for Colin Fraser's biography of that great man, Harry Ferguson. But while Ferguson's life and achievements are well documented, Philip Johnson's are not, and yet are regarded by many to have been of equal importance in different fields. This book, I hope, will redress the balance and introduce his work to a wider audience.

After Philip Johnson's death, the company continued to grow and prosper in the capable hands of other dedicated and innovative men. Without the help of several of these men, all pioneers and gentlemen whose loyalty and enthusiasm for the company is undiminished, this book could never have been written. Roadless was loyal to its employees and its customers, and it has never ceased to amaze me how much of that loyalty still exists, in return, for the company.

I have enjoyed writing all my books, but researching the Roadless story has been a special experience never to be forgotten.

Stuart Gibbard
February 1996

8

CHAPTER 1
From Swords to Plough Shares 1919 to 1930

Part of a consignment of several Morris Commercial half-tracks supplied by Roadless in 1926 to the Anglo-Persian Oil Company for exploration use in Iran.

It is impossible to write the full story of Roadless Traction without first looking at the life and philosophy of its founder, Lieutenant-Colonel Philip Johnson, who was managing director of the company for forty years.

Johnson was a man of great energy and outstanding commitment; he was an engineer of the highest ability whose active mind delighted in solving complex mechanical problems. More than this, Johnson was a visionary who saw his company and products as the solutions to removing many of the obstacles to the world's social and economic growth.

Philip Johnson was fascinated by technology and, although it has to be said that this fascination very occasionally led him to cast all practicalities aside, he had a unique talent for applying novel mechanical principles to the spheres of transportation and agriculture. Highly regarded in engineering circles, he was often invited to address the members of such worthy organisations as the Institute of Automobile Engineers, the Institute of Mechanical Engineers, the Royal United Service Institution and the British Association.

Convinced of the importance of technology in agriculture, Johnson fervently campaigned for the establishment of a professional body to further extend the scope of this field, leading to the setting-up of the Institution of Agricultural Engineers in 1938. The Institution made him an honorary member in 1947, and the Johnson medal was introduced and awarded to students who had achieved exceptionally high standards of merit.

For all his determination and intensity,

RIGHT:
The founder of Roadless Traction, Lieutenant-Colonel Philip Henry Johnson.

RIGHT:
Drawings of the cable suspension system taken from the patent specifications.

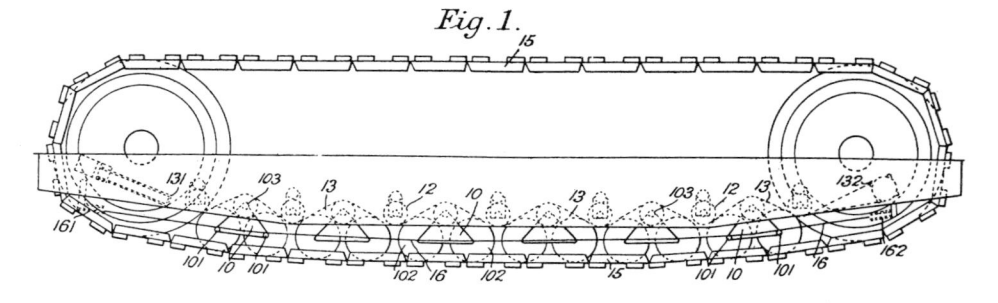

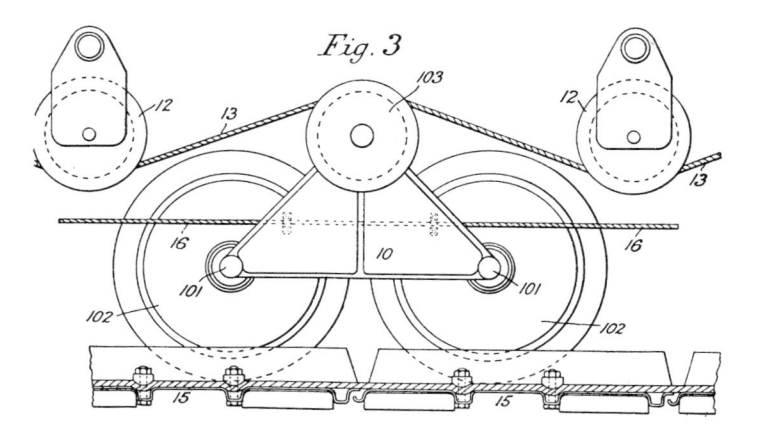

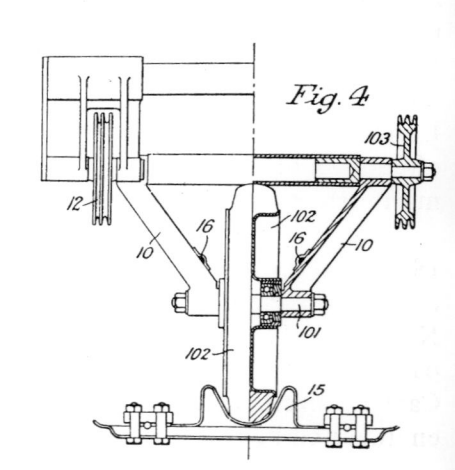

Johnson remained an astute and charismatic man of great charm throughout his eighty-eight years of life. He was meticulous in his work, and it has to be said that he did not suffer fools gladly. His company was eagerly sought by many keen to join his intelligent conversation and listen to him enthusiastically expound his theories of technical innovation. Even when relaxing, Johnson relished the opportunity to stretch the mind, and liked nothing better than to unwind over a challenging game of bridge.

Lieutenant-Colonel Philip Johnson was Roadless Traction – and Roadless Traction embodied all that Philip Johnson believed in and strove for.

Born in 1877, Philip Henry Johnson was educated at the King Edward VI School, Birmingham, before studying engineering at the Durham College of Science, Newcastle upon Tyne. He then embarked on a period of apprenticeship, working in the steel and shipbuilding industries of South Wales, and for a time joined Maudslay, Sons & Field of London, a celebrated engineering company often recognised as the finest manufacturers of marine steam engines in the nineteenth century.

Following the outbreak of the Boer War in 1899, Johnson tried to join the Army, but was rejected on the grounds of defective eyesight. Not to be defeated, he succeeded in working his passage to South Africa on a cattle boat. Once in Cape Town, his useful experience in steam engine engineering helped to get him seconded to the

ABOVE:
One of Philip Johnson's Medium D tanks, probably photographed in Woolwich.

LEFT:
The Medium D Star at Bovington, Dorset in 1930.

BELOW:
The Medium D Two-Star undergoing amphibious trials at the Experimental Bridging Establishment at Christchurch.

ABOVE:
The track arrangement for the experimental Light D, based on a 12 hp Overland car fitted with a light girder framework containing the cable suspension.

RIGHT:
Again based on an Overland car, the prototype Light D Star had tracks made from continuous rubber and fabric belts, and a different suspension arrangement.

RIGHT:
The Roadless Motorcycle, based on a Douglas machine fitted with cable suspension and tracks.

45th Steam Road Transport Company, Royal Engineers. This unit, initiated by Lord Roberts and under the command of Colonel J. L. B. Templer, employed steam traction engines to haul the heavy howitzers and field guns, and ammunition supply trains across the open African veld.

Incidentally, a small detachment of Electrical Engineers, with three Burrell engines and portable dynamos, under Lieutenant-Colonel Rookes Evelyn Bell Crompton, joined the 45th in Cape Town in 1900. Crompton was a kindred spirit to Philip Johnson. He had served in India as a Lieutenant in the Rifle Brigade from 1864 to 1868, during which time he was put in charge of the Regimental Workshops at Rawalpindi. Fascinated with motorised transport, he built his own steam carriage and experimented with a steam road train to replace the Viceroy's elephants. After leaving the army in 1875, he founded the famous Crompton Company supplying and installing electrical systems.

This period with the steam transport department in South Africa was to be a formative time for Philip Johnson: he gained first-hand experience of both the benefits and disadvantages of using mechanical transport in an off-road situation. It is not known whether Johnson met Crompton at this time, but if he did he could not have helped being influenced by the man's commitment to mechanised warfare.

Forty-six steam traction engines were eventually shipped out to the transport department in South Africa. Forty-one of these engines came from John Fowler & Company of Leeds, including four armoured Fowler B5s, four ploughing engines complete with ploughs to be used for forming earth fortifications and trenches, and a road roller.

After the Boer War ended in 1902, Johnson remained in South Africa for three or four years, and during this time he met his future wife, Mary Henrietta, and was married in Bloemfontein by December 1905. In 1906, he returned to England on Fowler's invitation to join the Leeds-based company. By this time, the sales of Fowlers' steam cultivating tackle were in decline on the home market, so the then managing director, Robert Henry Fowler, nephew of the founder, actively encouraged export sales in what was to prove a very successful attempt to get the Fowler name known worldwide. Johnson travelled widely on development work with the ploughing engines, eventually becoming the company's representative in India.

While in India, Johnson established offices in the York Building, Hornby Road, Bombay, and embarked on extensive sales tours as far as Madras and Eastern Assam. He was responsible for many successful sales of Fowler steam tractors and rollers, and was ably assisted by his wife who acted as his secretary. He also undertook to supervise the unloading of all the engines at the port, and very often personally delivered them to the remotest parts of the country under their own steam. This was a time during which Philip Johnson would acquire much of his knowledge of primitive agricultural methods – and a love of travelling.

Johnson returned to England with his wife and infant daughter, Carol, in August 1915, a year after the outbreak of the First World War, and was attached to the Ministry of Munitions. With his engineering background, Johnson soon became involved with tank development. In April 1916, he was given a temporary commission and was appointed second lieutenant in the Army Service Corps. The Army Service Corps personnel, notably the the No. 711 Company

TOP:
The White Staff Observation Car, which was fitted with cable suspension and tracks for the US Army by Johnson's Department of Tank Design and Experiment.

ABOVE:
Johnson took two Medium D tanks to India in December 1919 for trials which lasted six months.

RIGHT:
A Medium D tank undergoing tropical trials in India.

ASC, were provided as drivers to the Tank Corps, which was operating under the title of the Heavy Branch Machine Gun Corps to disguise its existence. Johnson was sent out to France on 20 September, five days after tanks were first used in the Battle of the Somme, to work under the chief engineer to the Tank Corps, Lieutenant-Colonel Frank Searle. Searle, who had been chief engineer to the omnibus manufacturers, Associated Equipment Company (AEC) of Walthamstow, had set up an impressive repair and maintenance unit at the Tank Corps headquarters at Bermicourt, a small chateau in France, not far from Agincourt.

By 1917, Searle had established extensive central workshops and stores on a thirteen-acre site of buildings and railway sidings at Teneur, two miles from Bermicourt. Johnson, by now a major, was put in charge of No.3 Advanced Workshops, the Tank Corps research unit, where he was ably assisted in his work by Sergeant-Major Charles William Skelton. Skelton, who was to become a life-long friend and associate of Philip Johnson, had commanded a tank during the Battle of the Somme. He had joined the Heavy Branch Machine Gun Corps after receiving a bad leg wound during a spell in the infantry.

Johnson worked on new track designs and developed a spring suspension system. He also

RIGHT:
A secondhand Crossley open tourer which was fitted with the cable suspension system by Roadless for customer, Mr Frank Reddaway.

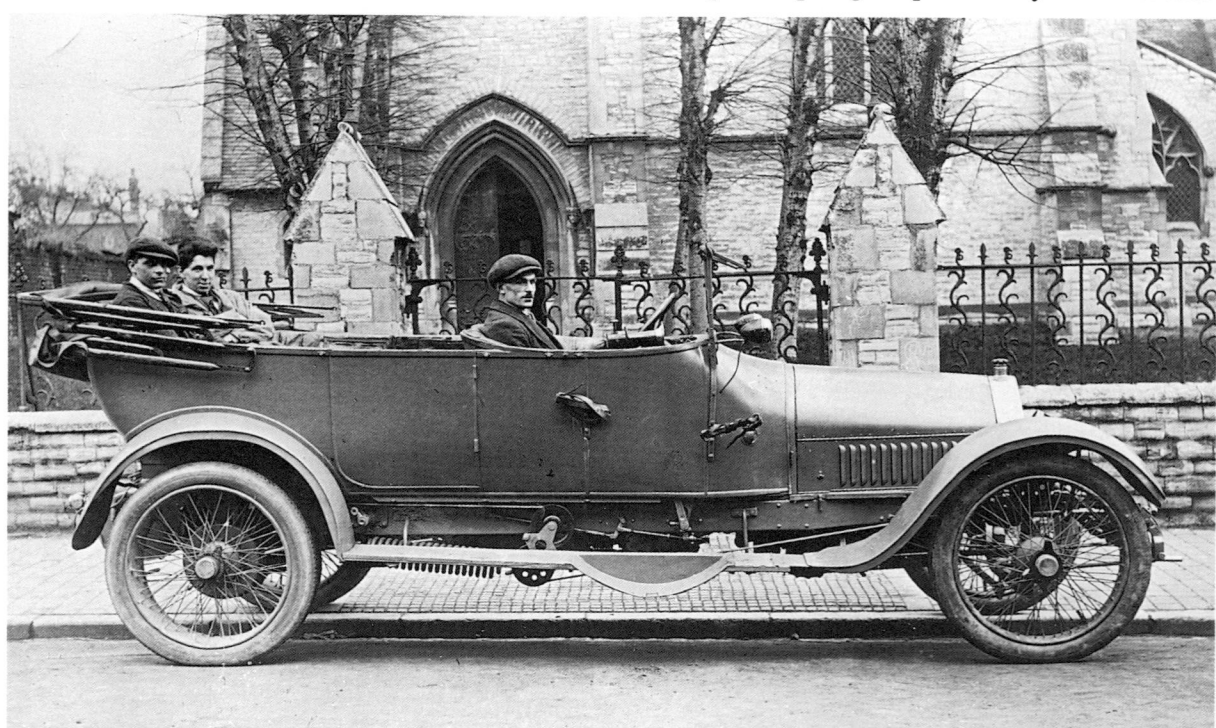

From Swords to Plough Shares

LEFT:
The Crossley with its wheels removed to show the cable suspension conversion which was fitted in 1920.

BELOW:
A Medium DM tank at Woolwich.

looked at ways of increasing the performance of the tanks by fitting larger engines and improving the transmissions. Taking his ideas a stage further, Johnson altered a Medium A Whippet tank to take his suspension system, and replaced its two 45 hp engines with a single Rolls Royce aero engine and an epicyclic transmission. The resulting vehicle had an unprecedented top speed of between 20 and 30 mph.

Johnson's developments soon came to the attention of the Tank Corps' general staff officer (GSO 1), Lieutenant-Colonel John Frederick Charles Fuller. Inspired by Johnson's work, Fuller, in early 1918, devised and drew up a new scheme of tank warfare tactics, later known as the 'Plan 1919', which hinged on the use of fast and manoeuvrable medium tanks to strike deep into the enemy lines. Fuller persuaded his fellow staff officers of the benefit of his plan, and Philip Johnson, now assistant chief mechanical engineer of the Tank Corps, was promoted to Lieutenant-Colonel and sent to England in May 1918 to the Mechanical Warfare Department with a brief to develop a new tank which became known as the Medium D.

The Mark V, the most successful tank used on the Western Front, weighed 30 tons fully laden and only had a range of twenty miles on a full tank of petrol at an average speed of 2 mph. The tracks needed overhauling after 150 miles, and operating costs worked out at £5,000 to £7,500 per 1,000 miles. Johnson's brief was to

ABOVE:
Probably photographed at the Tank Testing Section at Farnborough, this half-track based on a Fiat car was designed at Charlton Park for military use, but was the forerunner of the early Roadless machines.

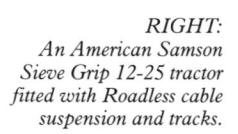

RIGHT:
An American Samson Sieve Grip 12-25 tractor fitted with Roadless cable suspension and tracks.

in September from Vickers. A wooden mock-up was ready by the time the Armistice was signed in November.

The main feature of Johnson's Medium D was a flexible track system which had wooden track shoes which were allowed to pivot laterally on swivel links joined together by steel wire. A suspension arrangement with pulleys kept under tension by wire cables and coil springs was also employed, and the complete system provided excellent weight distribution and allowed the vehicle to build a tank capable of operating at speeds of up to 20 mph with a 150 to 200 mile range, and using a more efficient low-maintenance track system.

Johnson divided his time between working at the Tank Corps Experimental Station, housed in the old McCurd lorry factory at Dollis Hill, Cricklewood, and frequent trips back to Bermicourt. Work continued on the Medium D; contracts were issued in August 1918 for four tanks to be built by Johnson's old company, Fowlers of Leeds, and another six were ordered manoeuvre easily at speed. Prototype tracks had been built up at Fowlers and fitted to a Mark V tank modified with the cable suspension system. This machine was demonstrated on 29 May 1919 to an assembled gathering of staff officers and other interested parties at Roundhay Park in Leeds, where it pirouetted alarmingly across the field, showering the spectators with discarded wooden track-plates, and reached unbelievable speeds (for a tank at that time) of up to 25 mph.

The first Medium D tank was completed by

Fowlers in March 1919, and was powered by a 240 hp Armstrong Siddeley Puma petrol aero engine driving through a three-speed epicyclic transmission. Sent to the Experimental Bridging Establishment at Christchurch for trials during June and July, it unfortunately caught fire and was irreparably damaged. Salvaged parts from it were used to complete the second machine, which went to Christchurch in August. The third and fourth machines from Fowlers were probably never completed.

The Vickers machines were built in the Wolseley factory in Birmingham. Two Medium Ds were completed, and two others were cancelled. Work also commenced on two new versions of the tank, the Medium D Star, with a four-speed epicyclic transmission, and the Medium D Two-Star with improved amphibious characteristics. All the Medium D tanks were supposed to be amphibious, but the Admiralty had declared them unstable, so the D Star and D Two-Star were made progressively wider to overcome the problem.

As far back as the autumn of 1918, Johnson had reached an agreement with the British War Office that any commercial value of the patents covering the suspension and track improvements would be retained by the officers involved in the original design work. To exploit these rights, Johnson registered Roadless Traction Limited on 14 March 1919, initially as a holding company to license the use of patents held by himself and his colleagues, Captain Edward Lionel Firth, Captain G. John Rackham, who had worked under Frank Searle at AEC, Captain Oscar Styles Penn, Lieutenant Shaw and Frederick Lamb. Johnson was managing director, and Firth, Rackham and Shaw, along with a Major Buddicom, were appointed to the board, which was

ABOVE:
The Samson Roadless tractor on demonstration in 1922.

LEFT:
The Samson Roadless tractor demonstrates its ability to cope with mud and ruts with ease.

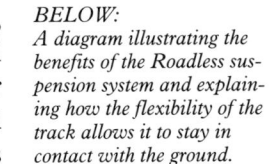

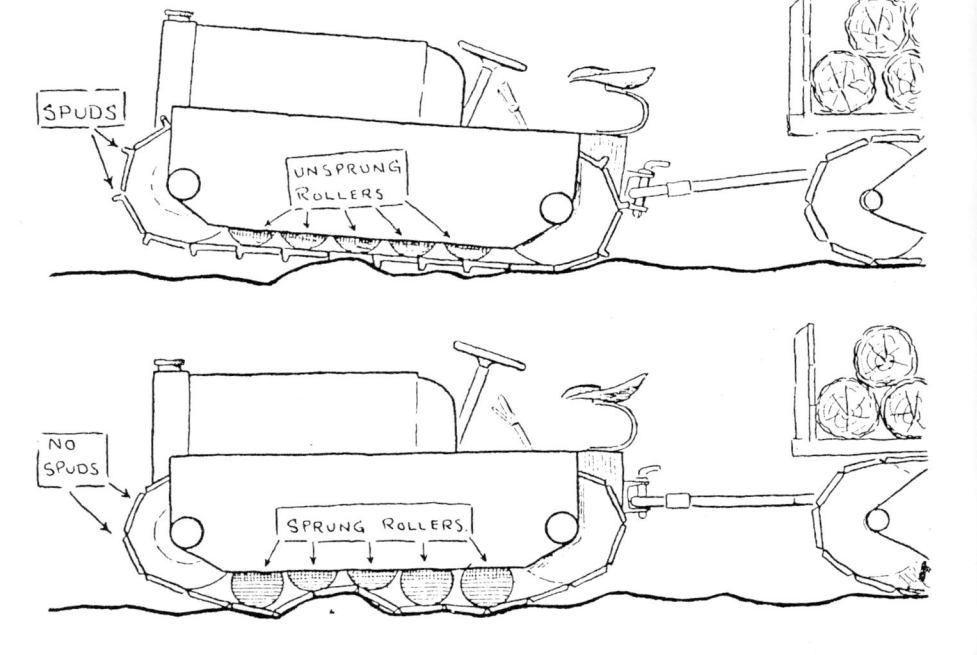

BELOW:
A diagram illustrating the benefits of the Roadless suspension system and explaining how the flexibility of the track allows it to stay in contact with the ground.

RIGHT:
The American Roadless Patents Holding Company licensed the Ordnance Department of the US Army to build this 16 ton army tractor in 1922.

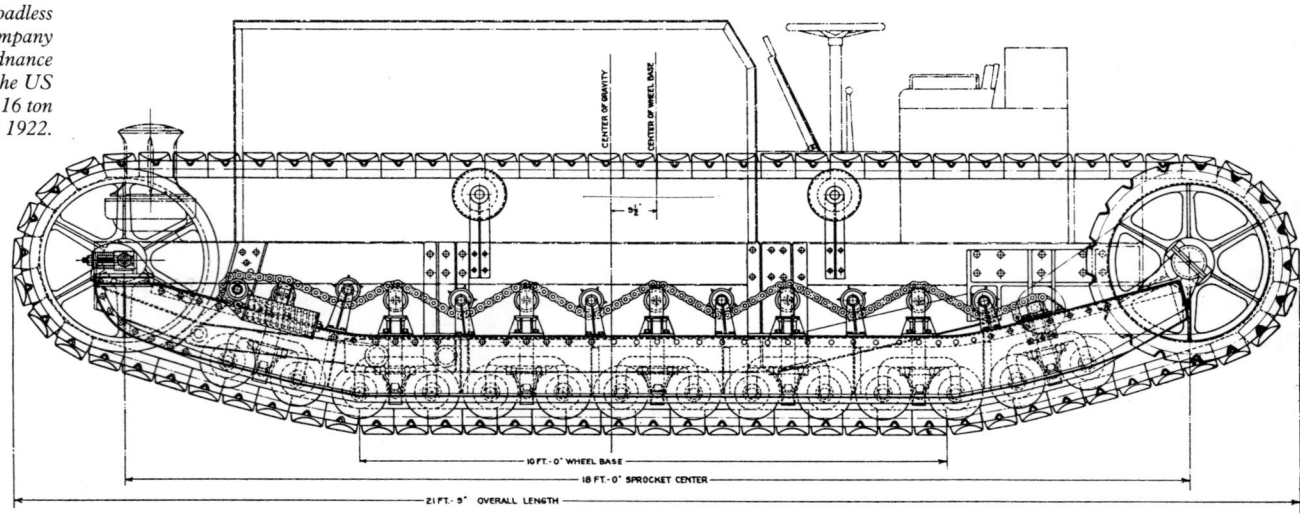

chaired by Lieutenant-Colonel Charles Willoughby Clark. The name 'Roadless Traction' was chosen to emphasise the commercial use of the track systems to provide 'traction independent of made-up or hard surfaces'. It was envisaged that the Roadless designs would eventually find applications all over the world.

During 1919, Johnson was made superintendent in charge of the Department of Tank Design and Experiment located at Dollis Hill, with living and office accommodation at the oddly named Maple Leaf Hut in Grosvenor Gardens. Johnson and his staff were involved in other projects apart from the Medium D tank, and were also working on developments for a light infantry tank. Unfortunately, the department was kept under-funded, and Johnson had to scrounge worn-out and surplus military vehicles from the Mechanical Warfare Department as test-beds for his experiments.

The Light D and the Light D Star were both based on American 12 hp Overland cars. The

RIGHT:
A Roadless Mack truck built under licence in the USA by the International Motor Company.

FACING PAGE BOTTOM:
The ideal conveyance after a few drinks on a Friday night – the Roadless Stretcher was based on two DW1 Orolo track units.

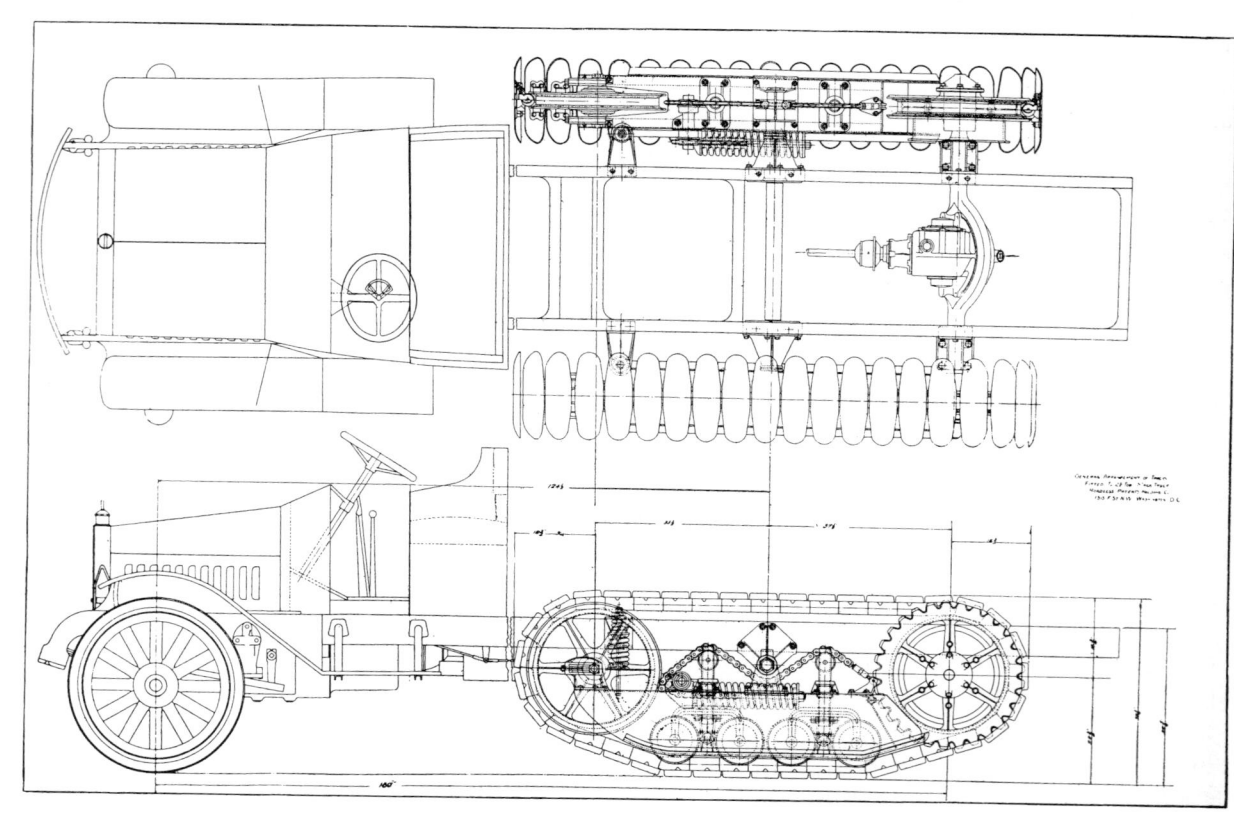

chassis of the Light D was fitted with a light girder framework on either side which contained the cable suspension, and a flexible track featuring large steel plates faced with rubber. The track used on the Light D Star consisted of a continuous belt made from rubber and fabric, and this vehicle had three pairs of drum-type rollers tensioned by the wire rope suspension. Both machines were capable of nearly 30 mph.

The rubber and fabric belt was also used on a Douglas motorcycle fitted with a scaled-down version of the cable suspension. At just under 3 hp, the machine would travel at 25 mph, and was designed for cross-country dispatch riders or recce work. It was often referred to as the Roadless Motorcycle.

Work was also progressing at Dollis Hill on a White Staff Observation Car fitted with an elaborate version of the cable suspension system and the rubber and fabric belt-type track for the United States Government. The vehicle was eventually completed in 1920 and sent to the Aberdeen proving ground in Maryland. Involvement with this machine had brought Johnson and his colleagues in contact with Major Lewis K. Davis. Davis, a consultant engineer based in New York, showed great interest in the Roadless systems, and persuaded the company to sell him an option on the sole patent rights for the designs in the USA for £100. An agreement was drawn up on 22 October 1919, giving Lewis twelve months to exercise the option for a payment of £10,000, due in instalments to Roadless Traction.

In December 1919, Johnson took two Medium D tanks to India for tropical trials to see how they performed under local conditions. The trip was not a great success; one tank failed with a broken clutch-shaft, and much time was wasted by futile attempts to protect the Medium Ds from the heat, either by covering them with asbestos, or hosing them down with copious quantities of water. The trials lasted over six months, and Johnson, who had taken

TOP:
Gunnersbury House in Hounslow, Middlesex – the headquarters of Roadless Traction from 1923.

ABOVE:
'By Jove, it's a wheelbarrow on tracks!' The Roadless Type G Barrow for gardens and sports grounds is exhibited at the Royal Show on the stand of John Allen & Sons (Oxford) Ltd.

*RIGHT:
A horse-drawn cart on Orolo tracks, built in 1928 by John Allen of Oxford for moving sand and stones in a gravel pit.*

*BELOW:
A flat trolley for single-bullock draught, manufactured in 1929 by the Khasia Sillimanite Company in India using Roadless DW1 Orolo track units.*

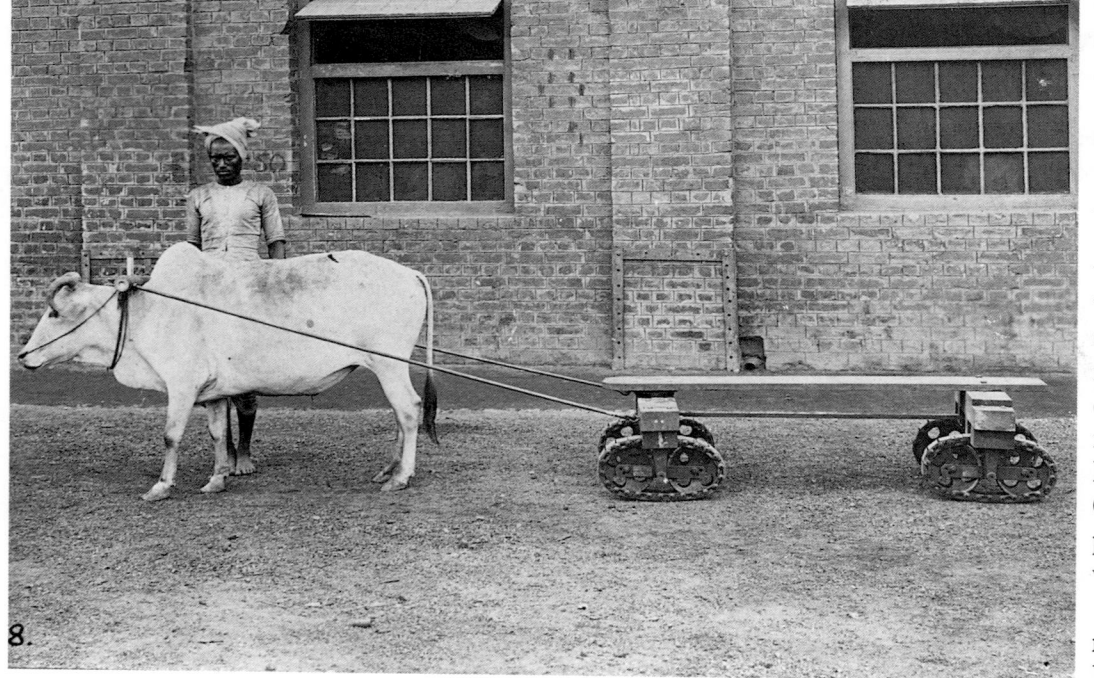

his wife and six-year old daughter with him, did not return until 12 June 1920. It appears that Johnson wore many hats, and was technically still in the employ of Fowlers, and made use of the visit to sort out some of the Indian affairs of the Leeds company. On his return, he handed over the post of Indian representative for Fowlers to Major H. S. Sayer.

Work at Dollis Hill carried on in Johnson's absence under the supervision of his assistant, Major Algar Abbiss Thompson. For some reason, Thompson, who had previously worked under Johnson in France, often sat in on the Roadless board meetings but never joined the company, and was eventually involved in the formation of a British subsidiary of the rival Citroen Kegresse firm who produced similar half-track vehicles.

Roadless Traction was managed for Johnson by Lieutenant-Colonel Clark from his home, aptly named 'Bermicourt', at Berkswell in Warwickshire. The secretarial work was carried out by a lady typist who went in for four to five hours a day, for which she received £1 a week. Most of Clark's time was spent trying to interest the motor industry in the cable system

From Swords to Plough Shares

LEFT:
The universally jointed and laterally flexible track that was commonly used on the early Roadless half-track vehicles.

as a means of improved suspension for wheeled vehicles, and details of the designs were publicised in *Autocar*. Clark was officially appointed as general manager of the company in March 1920.

The Dollis Hill establishment had two open-bodied Fiat cars on loan from the Army Service Corps, and one of these was fitted with the cable suspension system as a demonstration vehicle. It is interesting to note that this appears to have been more of a Roadless project than a military experiment. Johnson and his staff obviously had a free hand and managed to fit in a little commercial development on the side, but one can only speculate as to whether this had official sanction.

In late 1919, Clark arranged for the Fiat equipped with the cable suspension to be demonstrated in Coventry to the motor vehicle manufacturers, Mr. Bean of the Harper Bean Company, and Mr. Siddeley of Armstrong Siddeley. The car left Dollis Hill for Coventry on 20 December, with John Rackham at the wheel and Charles Skelton along as mechanic. However, the trip proved to be a fiasco – the wire-rope controlling the suspension broke several times, the demonstration was abandoned, and Rackham and Skelton only just made it back to London in time for Christmas.

The other Fiat was exchanged with the ASC for a Crossley Tender, which was found to be a better vehicle to take the cable suspension. Roadless's first customer for the suspension was a Mr. Frank Reddaway, and a second-hand Crossley Open Tourer was purchased for him and fitted with the cable suspension system.

During 1920, Clark unsuccessfully tried to sell Rover and Armstrong Siddeley the system. Interest was shown from the Wallace Farm Implement Company who wanted to produce a sprung version of its three-wheel Glasgow

TOP LEFT:
D8 Orolo track units, introduced in 1929 and capable of carrying 12 tons per pair.

ABOVE LEFT:
G2 Orolo track units, brought out in 1930 with improved rubber-jointed tracks.

21

tractor for road use, and an enquiry from AEC for the cable suspension to be fitted to a London General Omnibus. Unfortunately, neither vehicle was judged to be suitable to take the suspension system. Undeterred, Clark ordered himself a Standard motorcar fitted with cable suspension.

While Johnson was away, new premises had been found for the Department of Tank Design & Experiment at Charlton Park, Woolwich, and the move was made soon after his return in June. In 1920, Johnson and his staff were demobbed, but carried on their work at Charlton Park as civilians.

Plans to build seventy-five production models of the Medium D dropped to forty-five, and then twenty. Unfortunately, while the design suited the tactics of Fuller's 'Plan 1919', the machines were found to have many drawbacks for general service – not least their lack of reliability due mainly to the complex track suspension system. In the end, only two production models, known as the Medium DM (or D Modified), appeared. Built at the Royal Ordnance factory at Woolwich, they were sent to the Tank Testing Station at Farnborough. One later sank during amphibious trials in the River Thames.

Johnson's other work at Charlton Park included a tropical tank, powered by two 45 hp Tyler engines, for operating in India; small arms ammunition carriers using Ford 20 hp engines; a gun carrier, a supply tank, and a strange rigid rail machine, designed to have a low rolling resistance and driven by a Triumph motorcycle engine. The light infantry tank was completed and powered by an American Hall-Scott aero engine. It also had an improved universally jointed and laterally flexible track, known as the 'snake track', with lubricated and protected spherical joints. The Fiat car that had been used for

BELOW:
A Foden 6-ton wagon, the first steam vehicle to be fitted with Roadless tracks, is shown partly constructed.

RIGHT:
The Foden wagon was fitted with C-type Roadless tracks and demonstrated in June 1922 with a 5-ton load.

the cable suspension trials, was developed into a half-track, possibly for use as an armoured personnel carrier, and was fitted with the same track.

By July 1922, development costs for the Medium D had reached £290,000, and much more money would have to be spent before the machine was capable of going into full production. Unfortunately, the purse strings were held by the Master General of Ordnance's Department, who had already commissioned new tanks from Vickers. Johnson's Medium D project was scrapped, and the money was channelled into a new design of medium tank, created jointly by Vickers and the MGO's department, using several borrowed features.

The decision to drop the Medium D was probably more than a little political. If Johnson had been allowed to resolve the unreliability problems that beset his designs, the Medium D would without doubt have out-performed the unremarkable, although reasonably successful, Vickers Medium tank that replaced it. However it was not to be, as the Master General of Ordnance, Sir Noel Birch, who later joined the board at Vickers, regarded Johnson's independent department as a thorn in the side of the General Staff and an unnecessary expense. The decision to close Johnson's establishment was made in 1923, and Britain lost its Department of Tank Design – a move which was to leave the country lagging seriously behind in tank development at the outbreak of World War Two. For his work on tank design, Johnson had received the D.S.O., the C.B.E., and was made an officer of the Legion of Honour.

Johnson and his colleagues now had the time and opportunity to develop Roadless Traction into a viable commercial enterprise, involved not

LEFT:
The finished version of the Roadless Foden 6-ton wagon under test.

BELOW:
A 1925 Roadless conversion of the Foden D-Type steam tractor on improved B4 track units.

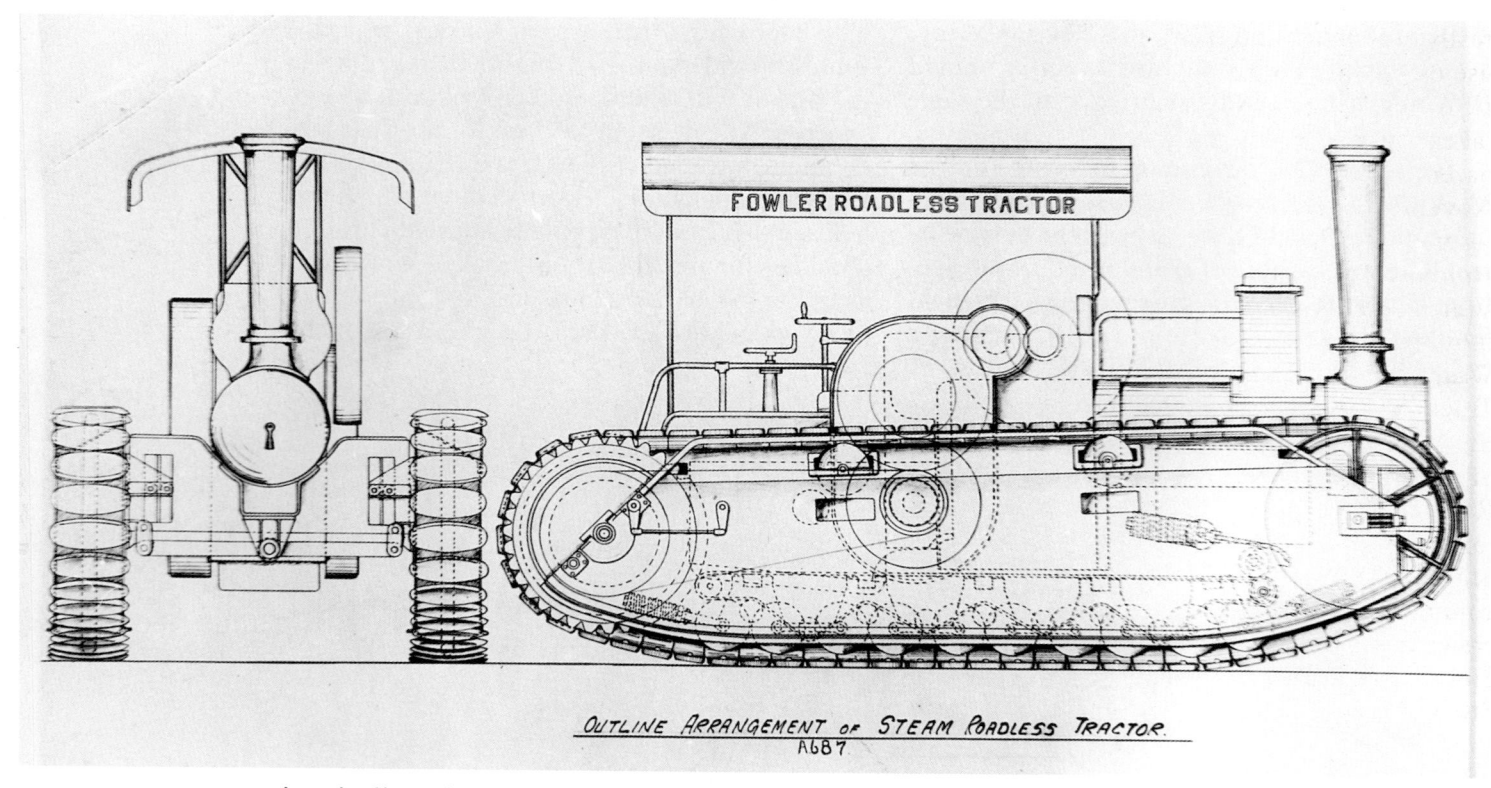

Roadless

OUTLINE ARRANGEMENT OF STEAM ROADLESS TRACTOR.
A687

ABOVE:
Drawings for the Fowler 'Snaketrac' which came out in 1924.

RIGHT:
The prototype Super-Sentinel Roadless tractor with B4 half-track units.

just in licensing, but also in design and development. The company was reorganised with Johnson and Clark as joint managing directors. Oscar Penn was made chief engineer, and Edward Firth became works manager, with Charles Skelton as his assistant. Annual salaries varied - Johnson got £1,000, Clark £800, Firth £650 and Skelton £270. Senior draughtsman, Frank Green was paid £300, and his assistant, Leonard William Tripp, got £146. The first drawing (for a B1 track arrangement) was made on 27 February 1923.

The company concentrated on what it knew best – designing track systems to fit various types of vehicles from wheel-barrows to a 100-ton road train. Prior to this time, Roadless tracks had only really been put to commercial use on motor vehicles in the USA. Major Lewis Davis had taken up his option on the patent rights, and in November 1921, he had formed the Roadless Patents Holding Company based in Wilmington, Delaware, later moving to offices in Washington DC. Roadless Traction received shares in the American company in lieu of money owed. John Rackham was persuaded to join Davis in the United States, and in 1922 the company demonstrated a crawler version of the Samson Sieve Grip 12-25 tractor from Stockton, California. Fitted with the cable suspension, this was a very advanced machine for its time, and exhibited excellent performance capabilities.

Licences to use the patents for the cable suspension and the snake track were sold to the Ordnance Department, US Army, for use both on a heavy tractor and an M1922 tank. The International Motor Company also bought rights to use the equipment on a half-track version of its 2½-ton Mack truck. However, by 1928 the agency for Roadless products had been taken over by the Isbell-Porter Company of Newark, New Jersey, and Rackham had returned to England to take up the position of chief engineer with his old company, AEC.

Back in England, Roadless Traction had built a couple of experimental vehicles, but desperately needed a new base to operate from. Premises were located in the form of Gunnersbury House and gardens, a former nunnery off the High Street in Hounslow, Middlesex, which Johnson bought off the trustees of the Little Company of Mary for £3,800 on 12 November 1923, and leased back to Roadless for £250 per year. The house had four floors, and was converted and modernised to provide both office space for the company, and living accommodation for Johnson and his family. Johnson's office was positioned on the ground floor, and at one time his secretary was inconveniently located in the attic. The drawing office, under Green and Tripp, was established in the basement which had been previously used by the nuns as a mortuary. All the drawings from this office were numbered, listed and dated in leather bound ledgers and then filed away in mahogany drawer chests. Very many of the original drawings and

ABOVE:
The Sentinel-Roadless undergoing trials in 1924.

LEFT:
The production model Sentinel-Roadless tractor. Most were exported to Africa.

RIGHT:
The experimental Roadless 'Waveless' road roller, built in 1926 by John Allen & Sons of Oxford and based on an old Aveling & Porter 10-ton steam roller.

BELOW:
The Waveless road roller was fitted with four heavy duty Orolo track units to exert pressure on the road without rippling the surface.

ledgers have still survived to this day in the same drawers.

At the outset, Roadless expected only to carry out small-scale manufacture, concentrating more on the initial design and development work for its products and leaving large-scale manufacture to its licensees. Rights to use the design patents were usually charged at 5 per cent of the finished value of the machine. Any manufacture and development work was carried out at the bottom of the two acre garden in an old stable block which was equipped with ex-army machine tools.

Before looking at Roadless's early products, it is necessary to understand the philosophy of the company. To Philip Johnson, the company was more than just a means to make money. Roadless was the personification of a vision shared by Johnson and many of his wartime friends that off-road traction was the solution to many of the world's social and economic problems.

Johnson's former mentor and army colleague, Colonel J. F. C. Fuller, took the dream one stage further in his book, *Pegasus*, published in 1925 and extolling the virtues of Roadless vehicles. In this small volume, he discusses the problems of over-population and unemployment facing post-First World War Britain, to which he offers two simple solutions – a coercive scheme of birth control, or to shift the population 'from our over-populated little island into our under-populated Dominions and Colonies'. The Government, he concedes, would not tolerate the first solution, although he adds that he is 'of opinion that it

LEFT:
Believed to be the first Roadless half-track motor-vehicle for civilian use, based on a Rover 12 hp Colonial chassis and built in July 1923.

BELOW:
An experimental Roadless half-track conversion of an AEC Y-Type 4-ton lorry chassis in 1926.

might well be made compulsory amongst politicians'.

The real solution, as Fuller saw it, was to open up the under-developed countries of the world and establish new industries in them, creating greater employment. To do this required an economic and reliable transport system. Laying new roads and railways across these vast countries was too expensive, and so the answer was to use the Roadless tracked vehicles that could travel along the existing rough cart tracks. Many of Fuller's beliefs were probably somewhat idealistic, but one of his concluding sentences echoes both Johnson's philosophy and the raison d'etre for Roadless – 'If the road will not suit the vehicle, the vehicle must be made to suit the road.'

Johnson's views were certainly broader than Fuller's, but he also felt the need to introduce the aims of his firm to a wider audience. Thus in July 1928, the first edition of *Roadless News*, a monthly news sheet outlining the objectives of the company, was published. *Roadless News* was regularly produced, although not always monthly, up to 1963 – its demise probably coinciding with the time when Philip Johnson would cease to be actively involved with the company.

Between 1921 and 1928, Roadless spent nearly £50,000 on developing its products – a sum of money which was a small fortune in the 1920s. To raise capital, debenture shares were issued, the bulk of which were taken up by three investors led by Sir William Jones, who became chairman of the company. Johnson also persuaded Cecil Booth, the inventor of the Goblin vacuum cleaner and founder of BVC, the British Vacuum Cleaner company, to invest in Roadless and become a director. Booth had

ABOVE:
A French Peugeot 4-ton truck fitted with Roadless tracks.

RIGHT:
This 2-ton Vulcan lorry was fitted with Roadless tracks and demonstrated on sand dunes near Southport, not far from its Crossens factory, in 1925.

BELOW:
This half-track was based on an Austin 20 hp taxi chassis and eventually sold to a Highland estate in Scotland for grouse shooting.

originally met Johnson after his return from the India tour, when he had tried to persuade BVC to develop a vacuum process for collecting tea leaves in Bombay.

One of Roadless's first and most commercially viable ventures was the introduction of the Orolo track units. Originally designed for the British Admiralty to transport a 6 in. naval gun over soft ground, the Orolo unit consisted of a self-contained bogie with rigid girder tracks running around two or three rollers. The track was designed to be self-locking so that it would form a predetermined radius to give it the same performance as an extremely large diameter wheel with very low rolling resistance.

The Orolo units were available in several different sizes, and could be used in place of the wheel in a variety of applications, including bullock and mule wagons for Africa, India and South America, log carriers and lifeboat carriages – and the Roadless drawing books even show details of an Excelsior washing machine so equipped in 1932, but for what purpose is not known.

When the Orolo unit was first conceived, a scaled-down version was tested on Philip Johnson's garden wheelbarrow. So successful was this unit, that Roadless decided to put the tracked wheelbarrow into production. A couple of hundred were made towards the end of 1927, and it became a popular addition to the company's early product line. At the other end of the scale were the D8 Orolo track units, introduced in 1929 and capable of carrying 12 tons per pair. The company eventually built units during the Second World War which would carry 20 tons on each track-bogie. The Orolo track units remained in production until well into the 1960s.

Much of Roadless's early work centred on developing off-road vehicles. Many half-track versions of several different steam and motor vehicles were built. Most never progressed beyond the prototype stage and very few went into full-time production. The tracks used on the early vehicles were a derivation of the original flexible snake track developed for the tanks. The advantage of the system was that the flexibility allowed the vehicle to turn easily without skidding sideways and also compensated for uneven ground, giving a smoother ride. C-type tracks using the cable suspension system were eventually replaced by the B-type employing an improved suspension system with rubber-tyred cast-steel rollers mounted on leaf springs.

A Foden 6-ton wagon, demonstrated in June 1922, appears to be the first steam vehicle to be fitted with Roadless tracks. Designed for load carrying, the machine used the early cable suspension track system without much success. A later conversion of the Foden D-Type steam tractor was brought out in 1925, using the improved roller and leaf-spring suspension, and was fitted with steering brakes to aid turning. Again, only one was built. The disadvantage of using the Foden was that it had a horizontal boiler which made it difficult to keep the water at the right level when climbing slopes. It is believed that both machines were eventually scrapped.

The Fowler 'Snaketrac' Roadless Tractor was

ABOVE:
A proposed design for a Wolseley car chassis fitted with Roadless tracks.

LEFT:
The Roadless half-track version of the 18 hp Guy 1-ton truck was built from 1923 onwards.

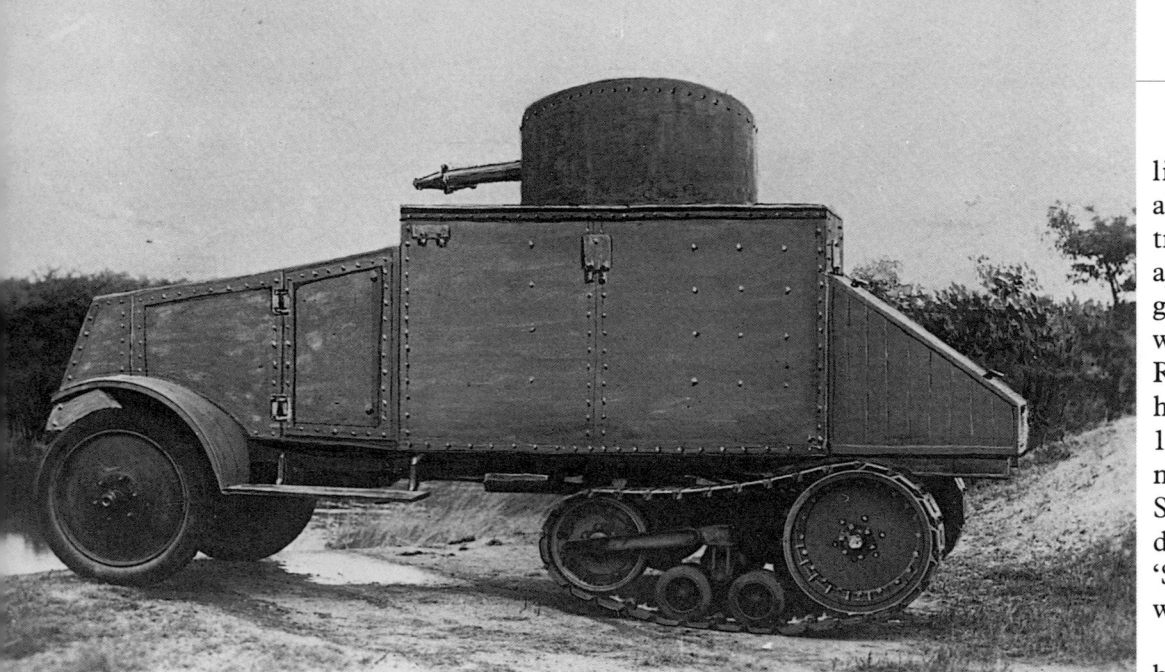

ABOVE: An artist's impression of the proposed Guy armoured car on Roadless tracks.

RIGHT: The Daimler CK 3-ton lorry was shown with Roadless tracks from 1924.

first exhibited at the 1924 British Empire Exhibition held at Wembley. This experimental full-track machine is believed to have been under construction as far back as 1919. Fitted with the snake track, it probably evolved out of the original work that the Leeds company carried out for Philip Johnson on tank development. Given the works number 14761, it was based on a Fowler A9 compound traction engine which was the only example to be built of a new range of A9 engines authorised in 1920. The machine was exhibited for sale following the signing of a licensing agreement in 1923 allowing Fowlers to use the tracks commercially. Too heavy and cumbersome to be of any great use, it was never sold and was probably cut up for scrap. Roadless drew up plans for a half-track Fowler tractor in late 1924, but it appears to have never left the drawing board. Some details also exist of Fowler designs for an oil-engined 'Snaketrac' in 1927, but again it was almost certainly not built.

Also in 1924, Roadless brought out a half-track version of the oil-fired Super-Sentinel steam tractor built by the Sentinel Wagon Works Limited of Shrewsbury. Designed for direct traction, this machine had two Roadless B4 half-track units in place of the rear wheels, and was fitted with heavy-duty Feredo brakes. The 75 hp vertically-boilered Super-Sentinel was powered by a horizontal twin-cylinder high-pressure engine. At its first outing for the press it gave an impressive demonstration travelling over boggy and deeply uneven terrain, climbing a 1 in 1 slope, and hauling a 58-ton train of derelict steam engines up a slight incline. The price for the machine was set at £1,250.

Some of the field trials for the Sentinel-Roadless were carried out on Bomford Bros' farm at Pitchill; Leslie and Douglas Bomford, both of the family who founded the famous Bomford & Evershed agricultural machinery company, were friends of Stephen Alley, Sentinel's managing director. Leslie was to join Sentinel and take responsibility for the steam tractor project, and accompanied the Sentinel-Roadless to Kenya where it was used for field cultivations with a gang of three five-furrow disc ploughs. About fifteen of the Sentinel half-tracks were built and exported to South and East Africa, where they operated with only limited success, proving to be slightly under-powered and suffering from rapid track wear. Some of the Sentinels were eventually fitted with the later rubber-jointed tracks, but this was no improvement as the tracks stretched under the torque of the engine, allowing the drive sprockets to slip.

Roadless's final flirtation with steam power came in the form of the 'Waveless' road rollers, jointly built with John Allen of Oxford. The first experimental machine, based on a redundant Aveling & Porter single-cylinder steam roller, was conceived in 1926. The conventional rollers were replaced with four heavy-duty Orolo track-units, one either side at the rear, and a pair mounted together in place of the front roller. The idea behind the machine was that the Orolo units would exert more pressure where it was needed on the road surface, and do away with the rippling effect sometimes caused by conventional rollers – hence the name, 'Waveless'.

An improved version of the roller was exhibited at the Public Works Roads and Transport Exhibition in November 1929. This machine was fitted with a vertical Sentinel boiler and twin-cylinder engine. Priced at £1,250, the roller was considered expensive for general road maintenance. In the end, one of the improved machines joined the original experimental roller on land reclamation work in Holland, and one further vertical-boilered machine went to Morecambe Town Council in 1930.

Roadless converted far more motor vehicles than steamers to half-tracks. The idea behind the commercial developments was probably born with the Fiat half-track that was developed back at Charlton Park. This was followed by what is considered to be Roadless's first civilian half-tracked motor-vehicle, based on a Rover 12 hp Colonial chassis, and built in July 1923. Later

TOP:
The Daimler Roadless was fitted with B3 track units and powered by a four-cylinder Knight sleeve-valve petrol engine.

ABOVE:
Roadless tracks fitted to an American designed FWD truck built by Four Wheel Drive Motors Ltd. of Slough.

ABOVE: An FWD truck on Roadless tracks hauling supplies in the Sudan.

vehicles include half-track versions of an AEC Y-Type 4-ton chassis, a Peugeot truck and an interesting conversion of a 2-ton chassis from the Vulcan Motor & Engineering Company, which was demonstrated on the beach near its Southport factory. Some half-track conversions were also based on light cars, including a vehicle designed for a highland estate using an Austin 20 hp taxi chassis. Drawings exist of half-tracks designed for Albion, Maudslay, Ford, Wolseley, Arrol Johnston and Somua vehicles, but photographs do not, so they were probably never built.

It seems that many of these early half-track conversions were evolved by Roadless, themselves, without any initial assistance from the companies on whose chassis they were built. After developing the vehicles, Roadless would approach the manufacturers involved in the hope that they would take up the designs, as some did including Guy, Daimler, FWD and Morris, otherwise the vehicles never went beyond prototype stage and disappeared.

Sydney Slater Guy, founder of Guy Motors Limited, based in Fallings Park,

Wolverhampton, had an affinity for the countryside and foresaw a need for trucks for the farmer and off-road use, and actively encouraged Roadless to convert his vehicles. A prototype 18 hp Guy truck on half-tracks, designed to carry one ton, was built in 1923, and a production version, based on the 2½-tonner, followed in 1924. Guy already had well established agencies in Cape Town and Sydney, and two of the Roadless-equipped trucks made an impressive 2,000 mile trek across Australia in 1925. Sketchy details exist of plans to put an armoured car version of the Guy half-track into production, but it is unlikely it was ever built. Several Roadless conversions of the Daimler CK 2- and 3-ton trucks were built between 1924 and 1925, and drawings for an armoured version of the Daimler Roadless lorry were also made.

Roadless probably gained the best performance from their half-tracks after fitting them to FWD lorries from 1926. These American designed four-wheel drive trucks had been built in England by Four Wheel Drive Motors Limited, based at Slough, since 1921. With the benefit of a driven front axle, the FWD trucks overcame the problem encountered by several of the Roadless half-track conversions of downward force exerted by the tracks pushing the front wheels into the ground. One Roadless equipped FWD was supplied for desert work in the Sudan. Working in excessively high temperatures on difficult terrain, it covered 2,000 miles without problem in the first three months. Several FWD Roadless tractors were supplied to

ABOVE:
Several interested parties are given a lift during a demonstration of the Morris Commercial 1-ton Roadless Lorry. Note the extended radiator header tank to aid cooling.

LEFT:
A Morris Commercial Roadless half-track is used for hauling sugar beet with a 30 cwt trailer.

FACING PAGE BOTTOM:
A Roadless FWD tractor supplied to the RNLI for lifeboat launching during the late 1920s, is shown after being fitted with improved DG tracks in 1945.

William Richard Morris, later Lord Nuffield, was the first British truck to go into mass production. It was built in Foundry Lane, Soho, Birmingham from early 1924 onwards, initially only as the T-type 1-tonner. Roadless had a prototype half-track version ready by October the following year. The Morris Commercial 1-ton Roadless Lorry, as it was known, was the first real breakthrough in the off-road vehicle market for the Hounslow company and became one of its best known early products. Many more of these vehicles were equipped with half-tracks than any of the other early Roadless conversions, and it proved popular at home and abroad.

The advantage of the Morris truck was that it was easily available, reasonably priced, simple and light in construction, yet well-made and rugged enough to stand the rigours of off-road work. The Morris-Roadless trucks had the advantage of a larger

ABOVE:
A Morris Commercial Roadless Lorry supplied to Peru by Alexander Eccles and Company is photographed after having climbed Chills Pass near Lima in April 1926.

the Royal National Lifeboat Institution for launching lifeboats from Orolo track-equipped carriages.

The Morris Commercial, the brainchild of

RIGHT:
An interesting Roadless half-track conversion of a Morris Commercial for use as a personnel carrier or charabanc.

From Swords to Plough Shares

radiator header tank to increase the cooling capacity of the vehicle when working under severe conditions.

Roadless's first real commercial success came with an order for several Morris Commercial half-tracks to be supplied to the Anglo-Persian Oil Company for exploration use in Iran, following the signing of the Anglo-Persian treaty in February 1926, which gave APOC a twenty-five year oil exploitation contract. Among other export sales included at least one Morris-Roadless sold through Alexander Eccles and Company to telegraph engineers working in Peru. One of this company's Roadless-equipped vehicles is recorded as having climbed Chills Pass, near Lima, in April 1926, when other tracked vehicles had failed.

Roadless's association with Morris led to an interesting diversification back into tank manufacture in the 1920s. Major Giffard Le Q. Martel, who had been in the Tank Corps during the First World War, believed the need existed for a light, one-man tank or 'tankette'. Coming from an engineering background, he set about building the machine himself in the garage of his house near Camberley. He used an old Maxwell car engine for power, a back-axle off a Ford truck, and a set of Roadless track units ordered from Philip Johnson. A rear axle with two wheels was added for steering and stability. Demonstrated to the War Office during the middle of 1925, it generated enough interest for Morris Motors to be given an order to build four more prototypes, followed by eight production

TOP:
An experimental chassis design for the Morris Martel tank.

ABOVE:
The Morris Martel seen fitted with Roadless tracks. The rear wheels were for steering and stability.

35

machines, in conjunction with Roadless Traction, who carried out most of the test work with the Army. Known as the Morris Martel Tank, it was not very successful, and suffered from stability problems. Martel went on to build a second version based on a Crossley, but with Kegresse tracks. The Morris Martel was designed so that it could be used without the armour as an agricultural tractor, although it is doubtful whether any were employed as such. One unarmoured version was tested by the Royal Marines at Fort Cumberland. A Roadless-equipped Thorneycroft Hathi gun tractor also saw service with the Royal Marines at one time.

By the late 1920s, the snake track design was beginning to show its limitations. The lubricated pin-joints appear to have attracted dust and grit, leading to rapid track wear. This first came to light on the Sentinel-Roadless tractors in Africa, and prompted Roadless to re-think their track designs. The eventual solution that the company arrived at was to replace the troublesome pin-joints with flexible joints made from rubber blocks.

The new track designs fitted to a field artillery tractor based on the Morris Commercial underwent initial trials in 1927 with the War Department. Two of these artillery tractors eventually saw military service with the India Stores Department in India, where they proved capable of hauling heavy gun equipment at speeds of up to 25 mph. Some FWD Roadless trucks were also fitted with the new tracks. This rubber-jointed track was to herald a new era in track design for Roadless Traction.

Taking the concept of off-road traction a stage further in 1930, Roadless introduced a specially reinforced vehicle tyre designed to run at only 10 psi for increased adhesion over rough or soft ground. The tyre was also capable of being run without any air in it at all – an idea far in advance of the Dunlop Denovo tyre of the 1970s.

One other early Roadless design worthy of a mention is the 100-ton road train, which unfortunately never got past the

BELOW:
Artist's impression of the Morris Martel one-man tank or 'tankette'.

RIGHT:
An unarmoured version of the Morris Martel undergoes trials with a trailer fitted with Orolo track units.

drawing board stage. Originally drawn-up in 1925, but not announced until 1929, the road train was to consist of a 7-ton crawler tractor, powered by a 300 to 400 hp slow-revving diesel engine, hauling four trailers, each of 25 tons capacity, mounted on Orolo track units. Designed for use in the under-developed countries, it was envisaged that it would only travel at 5 mph and would cope with crossing the most difficult terrain. Undoubtedly it was within Roadless's capabilities to build the train, but the company, probably sensibly, saw it was perhaps a feasible, but not financially viable, proposition.

With export sales seen as Roadless Traction's main market, a world-wide network of agencies was quickly built up – extending to well over 100 countries by 1930, stretching from Afghanistan to Antarctica. Principal among these distributive organisations were the firms of John Allen of Oxford, and H. C. Slingsby of London.

John Allen & Sons (Oxford) Limited acquired the sole manufacturing and distribution rights for Roadless track units throughout Great Britain and Ireland. Based in Cowley, the business had evolved out of the Oxfordshire Steam Ploughing Company which, founded in 1868, had later come under the control of John Allen. By the 1920s, the company was involved in agricultural and general engineering and were important manufacturers of road-making equipment.

Operating from London offices, the Slingsby company was formed in 1871 by inventor and designer, H. C. Slingsby. Slingsby, selling trucks, ladders, barrows and castors manufactured at its Bradford factory, was important to Roadless as the company had an already established network of overseas distributors, which by 1928 extended to more than twenty countries.

With business steadily expanding, Roadless decided to erect a new 150 ft by 30 ft, workshop at Gunnersbury House. This was completed by the end of 1929, and the company was poised ready to move into the 1930s with new designs for agricultural tractors on the drawing books.

ABOVE:
The Roadless Tyre introduced for cross country work in 1930. Designed to be used with very low air pressures, it would also run flat.

BELOW:
The design for the Roadless 100-ton road train as drawn up in 1925.

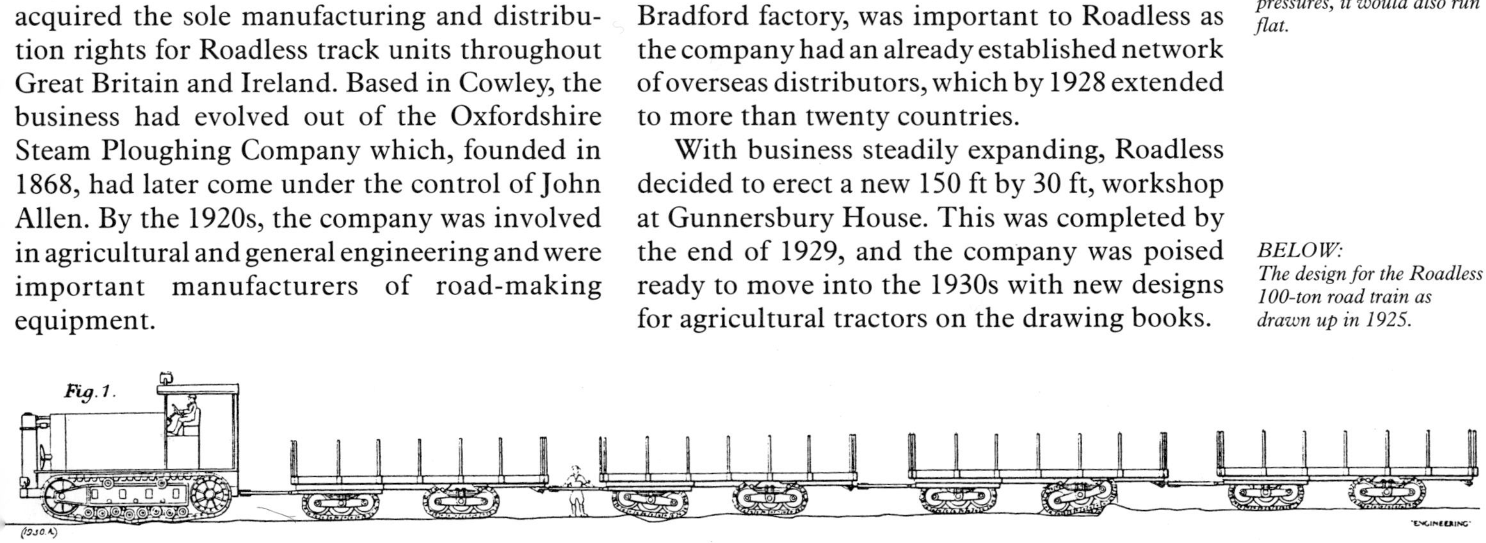

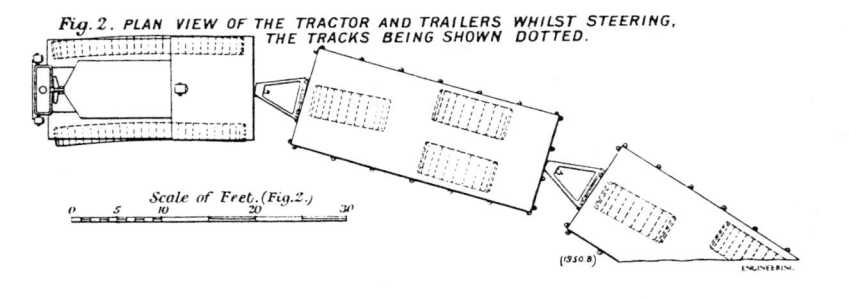

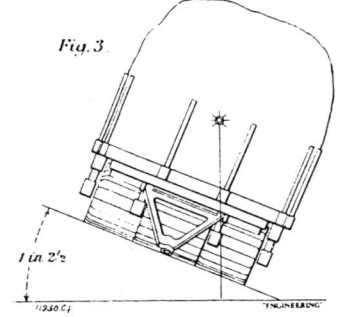

CHAPTER 2
Tractors on Tracks
1925 to 1938

A diesel-powered Garrett Roadless tractor makes light work of pulling a five-furrow Howard Colonial plough on the heavy loam of a Suffolk farm.

ABOVE:
The Peterbro Roadless tractor at work with a three-furrow plough.

RIGHT:
The Barford & Perkins Roadless tractor built at Peterborough.

Having enjoyed a certain amount of success in converting various steam and motor vehicles on to tracks, it was only natural that Roadless would start to investigate the possibilities of using its track equipment on the humble farm tractor. By coincidence, both of the first Roadless agricultural tractors were based on machines built at Peterborough.

The Peterbro tractor was built from 1920 to around 1930 by Peter Brotherhood Limited. It was powered by a four-cylinder petrol-paraffin engine of Ricardo design developing 30 hp. Several of the tractors were exported to Australia and New Zealand through Andrews and Bevan of Christchurch. The Peterbro Roadless tractor, fitted with B3 half-track units, was on Roadless's drawing books as early as 1925, but the machine was not announced by Peter Brotherhood until 1928. Details of the half-tracked model are sketchy, and how many were built is not known, but it may have been just the one example.

The second Roadless tractor was to come from another well-known engineering concern based in Peterborough - the road roller

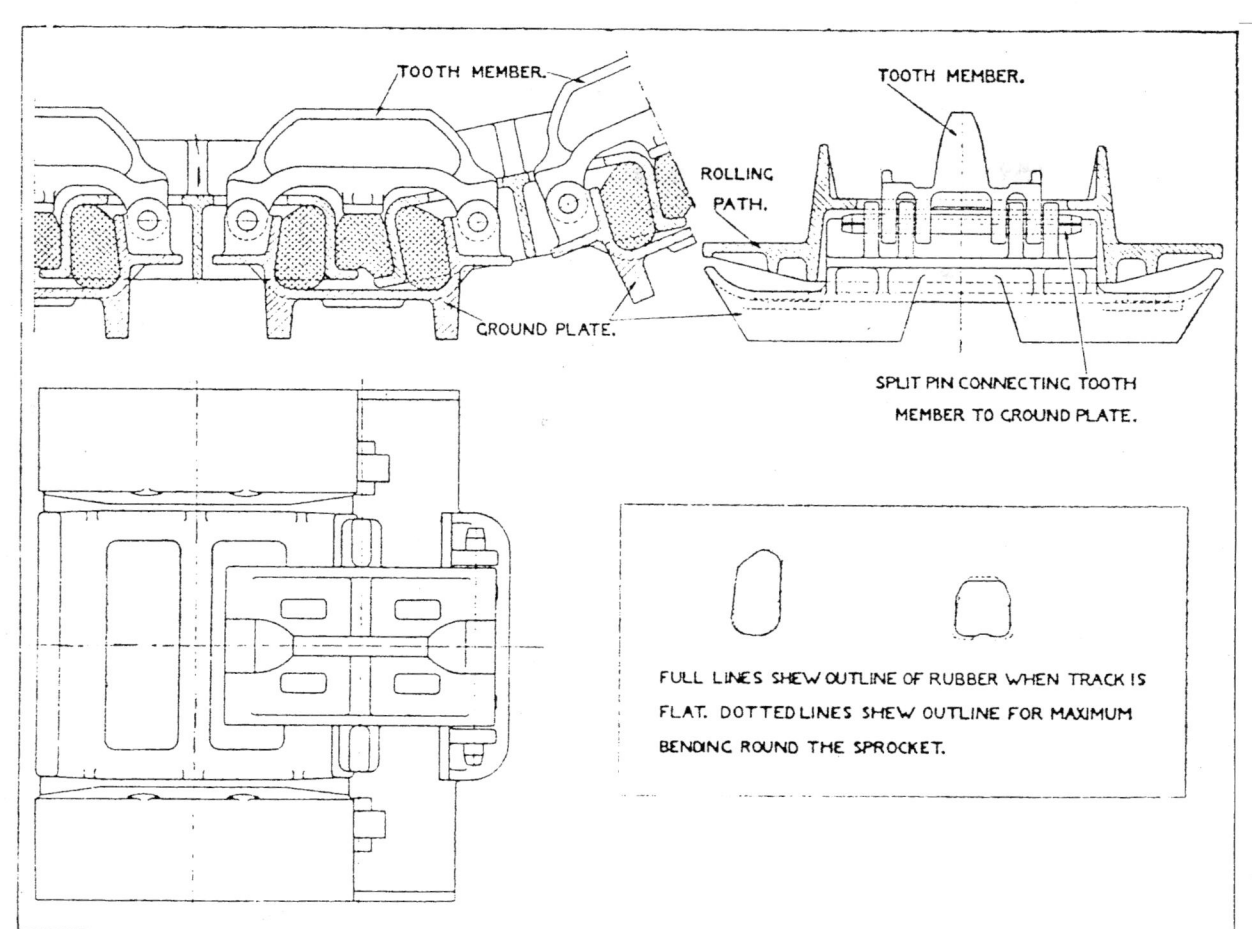

LEFT:
Detail of the rubber-jointed track.

BELOW:
The first Rushton Roadless tractor built in 1929.

manufacturers, Barford & Perkins Limited. The tractor was based on the Peterborough company's THD series road roller, powered by a rear-mounted vertical two-cylinder McLaren-Benz engine, supercharged on the Roadless to 50 bhp. The machine had three speeds in both forward and reverse. Steel road wheels were fitted in place of the front roller, with rubber-jointed Roadless E4 tracks on the rear. The machine could demonstrate a drawbar pull of 15,120 lb, but was over-complicated and too heavy at 11 tons to be of much use off-road, and so disappeared into obscurity.

It was the adoption of the rubber-jointed track system that was to allow Roadless to make its mark in the agricultural field. The company found the track easy to adapt to most makes of tractor, and the advantage to the farmer was the lack of maintenance required; there were no pins to wear out, deterioration of the rubber blocks was negligible, and no track adjustment or lubrication was needed. Also, the track was silent in operation. Roadless also claimed the frictionless joints led to higher efficiency, and that the rubber blocks eliminated shocks and led to greater adhesion. All the rubber-jointed tracks were given the prefix 'E' for 'Elastic Girder'.

Between the late 1920s and the period leading

RIGHT:
Two Rushton Roadless tractors which were dispatched to West Africa in late 1931

BELOW:
Three Rushton Roadless tractors supplied to Egypt for working on sand in 1932.

up to the Second World War, tractors on tracks were the mainstay of Roadless's business. It built many successful conversions on several different makes of tractor; all designed as full-tracks embodying the rubber-jointed type E3 tracks and skid-steered by Feredo-lined differential brakes. The first tractor of this type built was based on the Rushton.

The Rushton tractor was conceived and built by George Rushton while in the employ of AEC. Details of the tractor, initially known as the General and similar in design to the Fordson, were released in July 1928. Rushton Tractors Limited was formed early in 1929 as a subsidiary of AEC, and operated out of the AEC Works at Southall, Middlesex. The company's tractors, including an example on Roadless tracks, were first seen at the British Industries Fair held at Castle Bromwich, near Birmingham, from 18 February to 1 March 1929.

The company reorganised as Rushton Tractors (1929) Limited in late 1929. A lease was taken on AEC's Walthamstow factory and both the Rushton and the Rushton Roadless went into full-time production. The Rushton Roadless was available in two forms – a standard version with two bottom-rollers, and an extended version with three rollers. The slightly heavier extended version was more popular as it was more stable than the standard machine. The Roadless tractor was rated at 14-28 hp and cost £434 in extended form.

One of the first Rushton Roadless tractors built underwent trials in Russia over the 1929/30 winter. Several were sold at home and abroad, and the tractor was successfully demonstrated at the 1930 World Agricultural Tractor Trials held at Wallingford near Oxford. Unfortunately, the export successes of the Rushton Roadless led to the downfall of the Rushton company. In 1930,

Rushton received an order from France for a hundred Roadless tractors to be used in Algerian vineyards. The tractors were completed and shipped, but the customer defaulted on payment and Rushton went into liquidation.

Rushton was purchased by Frank Standen of F. A. Standen & Sons Limited. Standen set up a new concern, the Agricultural and Industrial Tractor Company, which operated out of the old Crown Hotel in St. Ives, Huntingdonshire, selling Rushton parts and a number of slightly improved complete tractors, now rated at 30 hp. It appears that many of the machines from here were supplied as skid-units for Roadless to fit with tracks. Sales were handled through Tractors (London) Limited, Broad Lane, Tottenham. Rushton Roadless crawlers were produced in to the mid-1930s.

Considering the popularity of the Fordson tractor in Britain, it is surprising that no Roadless tracked conversions of the machine were actively considered until the company was approached to build one for a customer in 1929. That approach came when the Invicta Motor

TOP:
Rushton Roadless tractor fitted with an Allen-Oxford backfiller attachment manufactured by John Allen & Sons from 1930.

ABOVE:
The first Fordson Roadless tractor on public demonstration hauling 3 tons of wet seaweed off Margate beach in April 1930.

LEFT:
This photograph clearly shows the steering brake arrangement for the Fordson Roadless tractor.

43

Engineering Works of Canterbury were asked by Margate Corporation to tender for the supply of a tractor to haul seaweed off the beach, a task previously undertaken by horses. After a normal wheeled tractor was found to be totally unsuitable for pulling the wagons heavily laden with wet seaweed on soft, loose sand, Invicta Motors looked into the possibilities of using a machine fitted with the rubber-jointed Roadless tracks that they had seen perform so well on Rushton tractors. As Fordson dealers, Invicta obviously wanted the Roadless crawler built on a Fordson Model N tractor skid-unit.

Plans were drawn up by Roadless in 1929, and the finished machine was demonstrated successfully in March 1930, easily pulling a 10-ton load of gravel on wet sand despite the trailer wheels sinking to a depth of 9 in. in places. A public demonstration was arranged the following month on Margate beach, and the Fordson Roadless showed it was capable of doing the job for which it was built – hauling a 3-ton wagonload of wet seaweed off the beach in front of an assembled audience of press and onlookers.

The introduction of the Fordson Roadless was well received, and before long enquiries for the tractor were flooding in from home and abroad. Like the Rushton, Roadless offered the Fordson both in standard and extended track form. By

RIGHT:
A Fordson Roadless at work cutting drainage channels on moorland.

BELOW:
Believed to be the second Fordson Roadless sold, this extended tractor was fitted with 30 in. wooden swamp shoes for operating on peat in Scotland, and is seen pulling a mole plough to a depth of 12 in.

September, one had been supplied on special swamp tracks for peat harvesting in Scotland, and others had gone out for demonstrations in the USA, Spain and Turkey.

The Model N Fordson, which had up to now been made in Cork, Ireland, was produced at Ford's new Dagenham plant from 1931. Roadless announced that an agreement had been reached with the Ford Motor Company who had approved the Roadless equipment for use with its tractor. Prices were set at £350 for the standard model, and £385 for the extended crawler. The Roadless equipment could be bought separately to the tractor from £180. This was the start of a long association for Roadless Traction with Ford and Fordson tractors that continued into the 1980s.

ABOVE:
This photograph demonstrates that even a Fordson Roadless can get stuck.

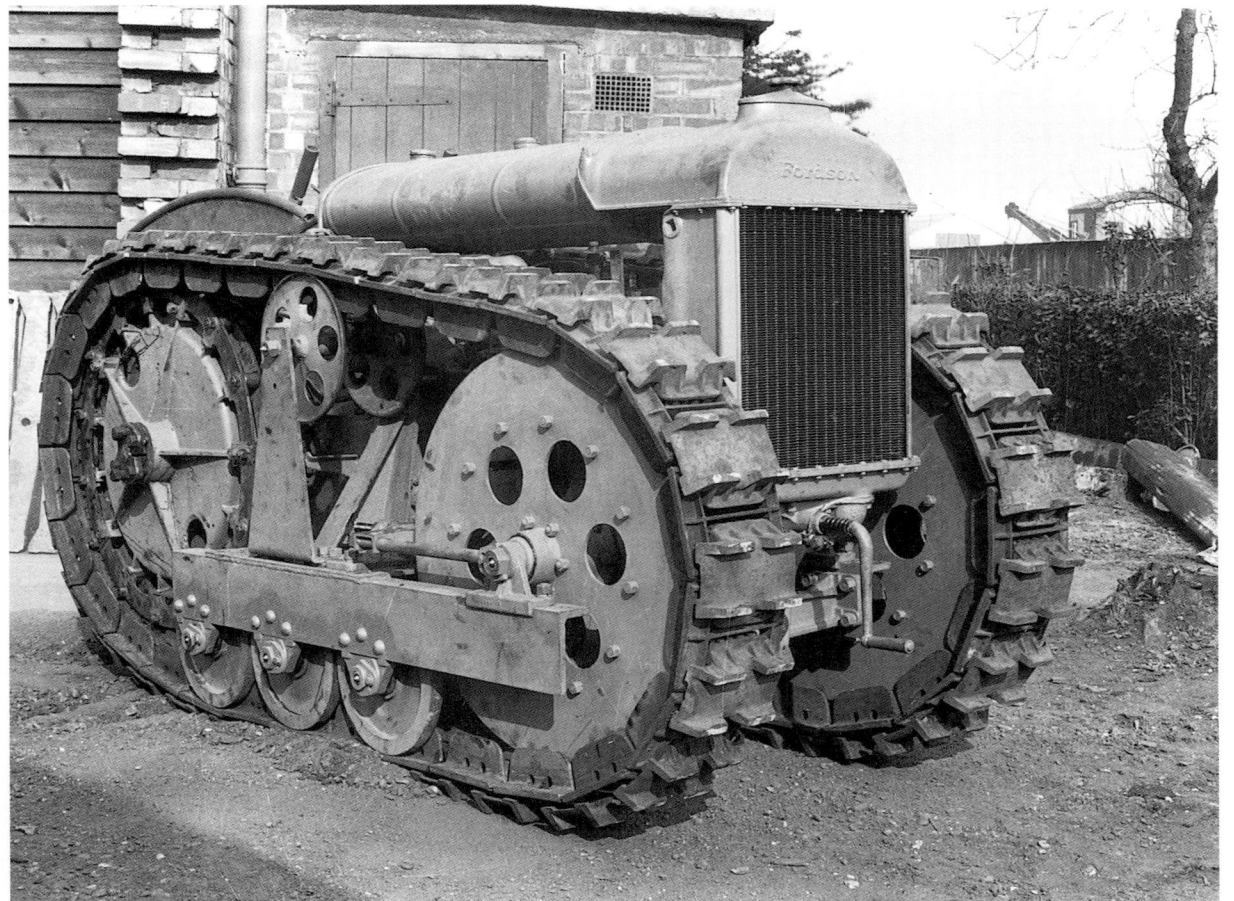

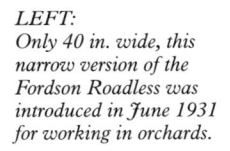

LEFT:
Only 40 in. wide, this narrow version of the Fordson Roadless was introduced in June 1931 for working in orchards.

The Fordson Roadless became the Hounslow company's best known prewar crawler and was produced well into the Second World War. A narrow orchard model was introduced in mid-1931, and the crawler was also used as the basis for other companies' machines, including a Canadian Wehr grader. Backfiller attachments for the Fordson and Rushton Roadless tractors were made by John Allen of Oxford, and, in 1937, one unique Roadless crawler, based on a Fordson fitted with an Ailsa Craig diesel engine conversion, was demonstrated as the ACR (Ailsa Craig Roadless). Even after the war, production of the Roadless crawler continued on the new models of Fordson tractor.

Following arrangements made with the

ABOVE:
Fordson Roadless supplied in 1938, and fitted with an Allen-Oxford backfiller blade for work on sea defences bulldozing sand and shingle.

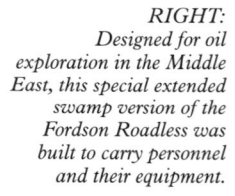

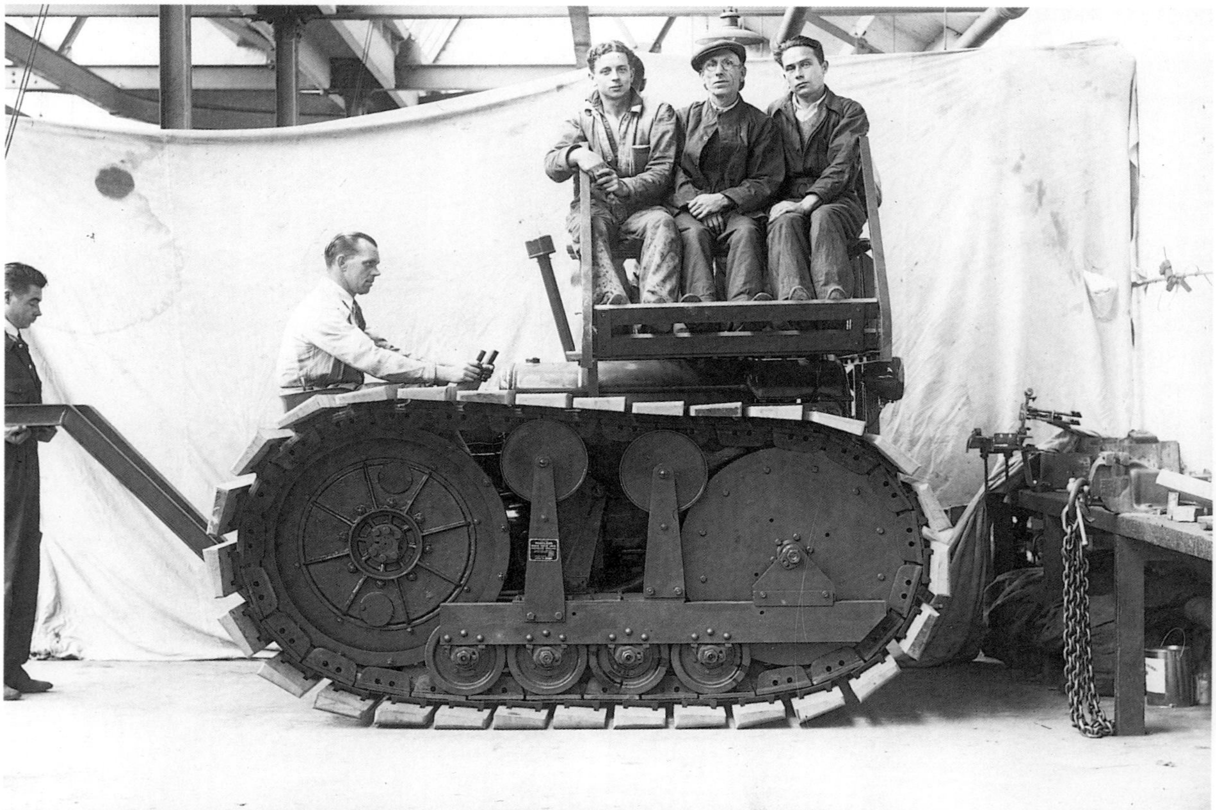

RIGHT:
Designed for oil exploration in the Middle East, this special extended swamp version of the Fordson Roadless was built to carry personnel and their equipment.

in that the steering wheel was retained to operate the steering brakes instead of the levers used on other versions.

The Case Model L Roadless, registering nearly 49 hp at the drawbar, and weighing in at around 4½ tons, was a useful size of tractor that was to prove popular with the industrial user. Two of these tractors were sold to the Iraq Petroleum Company in July 1931 to carry 20-ton loads of oil pipelines over desert country.

British importers, the Associated Manufacturers' Company (London) Limited of Kings Cross, the American Case, built in Racine, Wisconsin, was the third make of tractor to be converted on to Roadless tracks in any great number. The Model C was the first Case tractor to gain the benefit of Roadless equipment; it made its appearance in July 1931, rated at 29 hp and weighing 3 tons. Its bigger brother, the Model L Case, came out on E3 rubber-jointed tracks the following spring. The Case differed slightly from the previous Roadless conversions

Roadless also supplied the company with two bolster trailers, fitted with Orolo track units, to carry the pipes.

A special extended high clearance version of the Case L crawler was introduced in November 1935 for forestry work. This tractor had four bottom-rollers and 19 in. ground clearance. A further interesting variation of the Case Roadless appeared as late as 1940. Designed for the Turkish Government, this machine was based on the Case LH skid-unit fitted with a Hesselman low-compression diesel engine which relied on spark ignition.

LEFT:
A 1935 Fordson Roadless featuring improved tracks and track frames.

The Roadless Lifeboat tractor was a spin-off from the Case conversions. Roadless had been involved with the Royal National Lifeboat Institution since the 1920s, supplying Orolo track units for the lifeboat carriages, as well as track equipment for the FWD launching tractors, some of which had been later converted on to rubber-jointed tracks. When the RNLI wanted an improved machine for hauling and launching lifeboats, Roadless was the obvious choice to carry out the development work.

Brought out in 1936, the new lifeboat tractor was based on the Case L crawler, extensively modified and water-proofed so that it would work virtually submerged. All items on the tractor that would be affected by water were sealed. The spark plugs, generator and magneto were enclosed in a watertight aluminium casing. The air-breather and exhaust were extended to 7 ft 6 in. above the ground, and the tractor was fitted with four bottom-rollers and extra-wide 18 in. tracks. Several teething problems were encountered, but the first machine went to work with the Skegness lifeboat

BELOW:
The ACR tractor, a Fordson Roadless fitted with an Ailsa Craig two-cylinder diesel engine in 1937.

Roadless

*RIGHT:
The Case Roadless Model C tractor introduced in July 1931.*

*RIGHT:
A pair of Case Roadless Model L tractors supplied with bolster trailers for hauling 20-ton loads of oil pipelines in Iraq*

*BELOW:
A Case Roadless Model C backfilling with an Allen-Oxford angledozer.*

in May 1937. Five more were supplied to the RNLI during the next year, and one went to Holland for use on Terschelling island.

The Canadian Massey Harris was another imported North American tractor to be fitted with Roadless tracks. The Roadless version of the 25/40 hp was introduced in November 1932, and a smaller model based on the 12/20 hp followed in December. Like the Case, the steering brakes on both tractors were operated by the steering wheel, and the 12/20 hp was available in standard or extended form.

Towards the end of 1934, Roadless brought out improved heavy-duty versions of the Massey Harris 25/40 hp and Case L tractors. These new models featured several modifications including strengthened frameworks and 16 in. tracks. Also during the 1930s, Roadless converted two British-built diesel tractors on to tracks. The first of these was announced in December 1931, and based on the ill-fated Garrett tractor which never really got the chance to get off the ground.

The Garrett tractor was built by the famous steam engine manufacturers, Richard Garrett & Sons Limited of Leiston, Suffolk, who were part of the Agricultural and General Engineers (AGE) group. The Garrett Roadless was fitted with an Aveling & Porter Invicta diesel engine, developing 42/60 hp, and had standard E3 tracks. Initial trials proved it to be a powerful and economical tractor capable of a high work rate. It was also demonstrated to the Army with favourable results.

Unfortunately, only two machines were completed on

ABOVE:
A Case Roadless Model C fitted with improved tracks and strengthened track frames which were introduced in 1935.

LEFT:
A special extended version of the Case Roadless Model L introduced in 1935 for forestry work.

*RIGHT:
Fitted with a Hesselman diesel engine, this Case LH on Roadless tracks was supplied to Turkey in 1940.*

*RIGHT:
The Roadless Lifeboat tractor based on a modified and waterproofed Case Model L*

LEFT:
The Case Roadless Lifeboat tractor is used for a practise launch at Wells in Norfolk.

BELOW:
The first Massey Harris Roadless tractor, based on a 25/40 hp model, and introduced in November 1932.

BOTTOM:
Massey Harris Roadless 25/40 hp on heavy duty Roadless tracks.

Roadless tracks before the AGE group collapsed in February 1932, and the Garrett works were closed. Three more tracked versions were completed during 1932 by the receivers. These went to the Crown Agents for the Colonies, and were shipped to Gambia and Fiji. The Fiji tractor was fitted with a hydraulic bulldozer. The second of the two original Roadless tractors was fitted with a Blackstone diesel engine, and was not sold until March 1933 when it went to Malaya.

The Garrett concern was eventually bought by the Beyer Peacock company. In February 1933, Roadless announced that Garrett crawlers were again available, but now fitted with Gardner 4L2 diesel engines developing 38/45 hp. It is believed that only three more Garrett Roadless tractors were built, and were sold in April 1934 for peat extraction in Scotland.

Roadless's second attempt to build a British diesel-engined crawler tractor did not meet with much more success, although it was based on a better-known make of tractor – the Marshall. The first Marshall Roadless, fitted with standard E3 tracks, was announced in February 1933 and based on the single-cylinder 18/30 tractor. The first two machines built went to John Allen of Oxford, as did possibly four others. Another tracked 18/30 went to Robert Crawford of Frithville in March 1934, and Roadless built up one more on heavy-duty tracks in September 1934 for use in East Africa.

Roadless was also involved with a third British diesel crawler tractor, the Ransomes & Rapier RT50 brought out in January 1934. Unlike other Roadless crawlers, this was not a conversion of a wheeled tractor but was designed from the outset as a tracked machine using Roadless tracks. Produced by Ransomes and

RIGHT:
The Garrett Roadless tractor.

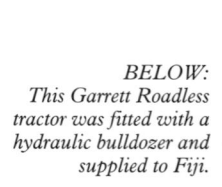

BELOW:
This Garrett Roadless tractor was fitted with a hydraulic bulldozer and supplied to Fiji.

BOTTOM:
Artist's impression of the Marshall Roadless 18/30 tractor.

Rapier Limited of Ipswich, this 8-ton crawler was powered by a 65 bhp Dorman Ricardo diesel engine. The tractor was fitted with a six-speed gearbox and multi-plate steering clutches. The tractor was too expensive to sell easily, and production was short lived. Ransomes and Rapier became better known for mobile crane manufacture.

Roadless built a crawler version of the Hungarian Mavag tractor in late 1931, and arrangements were made for this to go into production at the Royal Hungarian State Iron, Steel and Machine Works in Budapest. During this period, Roadless also produced several designs for experimental crawlers based on Austin, McLaren-Benz, Lanz Bulldog, Bolinder Munktell and Allis Chalmers tractors, but most never got beyond the drawing board stage.

Roadless tracks were also used on a prototype Rover tractor. This small, compact crawler was initiated by Philip Johnson's old First World War colleague, Colonel Frank Searle, who was managing director of the Rover Car company from 1929 to 1931. It was built in 1931 at Searle's home, Braunston Hall, and used a 839 cc twin-cylinder air-cooled engine designed for the prototype Rover Scarab car. Rumour has it that the machine evolved out of a desire by the Rover company to

LEFT:
The Ransomes & Rapier crawler at work.

LEFT:
Roadless version of the Hungarian Mavag tractor built in late 1931.

break into the agricultural market, possibly using Harry Ferguson's patents and incorporating his three-point linkage system. There is evidence of the tractor being fitted with a Ferguson plough, but the machine was found to be too unstable in the furrow and was scrapped at the end of 1931.

The Rover tractor was strangely similar in design to the future Bristol crawler, even down to the tiller steering. The Bristol tractor had started out as a Roadless project, and not just the tracks, but the complete machine, with the exception of the engine, was designed at Hounslow. As far back as 1930 the company had the notion to build a small, lightweight crawler tractor, powered by an air-cooled engine, for use on small farms and fruit plantations. Whether the Rover tractor was the result of collaboration between the two companies, and was the first prototype for the Bristol is an interesting thought, but pure conjecture.

What is known is that by 1932, after two years of design and experimentation, Roadless had completed two hand-built demonstration models of a lightweight crawler on rubber-jointed tracks. These prototypes were fitted with Douglas flat-twin engines, one a 10 hp overhead-valve water-cooled unit, and the other a new air-cooled 12 hp engine that would be used in the production model. The engine suppliers,

Douglas Motors, were looking for a new product to make up for falling motorcycle sales in the depression and were keen to take on the manufacture of the new crawler. Squabbles in the Douglas family had led to the company being purchased by a group of London financiers and reorganised as Douglas Motors (1932) Limited. An agreement was reached in June 1932 for the crawler to be built in Douglas's factory in Hanham Road, Kingswood, on the outskirts of Bristol from where the new tractor took its name.

Sales were to be handled by the Bristol Tractor Company of London under Walter Hill, formerly of Muir-Hill and Rushton, and on the board of directors of Roadless since Rushton Tractors collapsed in 1930. Details of the Bristol were announced in August 1932, and the price for the tractor was set at £155. Unfortunately the new Douglas concern ran into financial difficulties and went into receivership in November before any tractors could be completed.

Bristol Tractors Limited, with Walter Hill as managing director, was formed on 5 April 1933 to manufacture the Bristol from a factory in Sunbeam Road, Willesden. The agreement with Roadless was that Bristol Tractors would pay £1,000 for the design, and then a royalty of £2 per tractor with a minimum of £500 per year. The first sixty tractors were produced with V-twin British Anzani engines. This power-unit proved unreliable and was phased out by the end of the year. The crawler was then offered with the choice of a Coventry Victor diesel, or a water-cooled flat-twin supplied by the Jowett

RIGHT:
The 1931 prototype tractor produced by the Rover Car company and fitted with a Ferguson plough.

RIGHT:
The prototype Bristol crawler fitted with the Douglas flat-twin air-cooled engine.

ABOVE:
Bristol Tractors' production model crawler on demonstration in January 1934 at Heston near Hounslow. This tractor has the Anzani V-twin engine.

LEFT:
A later Bristol crawler powered by a water-cooled Jowett engine in 1934.

BELOW:
Bristol crawler fitted with the flat-twin Jowett engine and an experimental Roadless track arrangement.

car company.

Bristol Tractors also ran into financial difficulties, partly because of the minimum agreed royalty it had to pay Roadless, and partly because problems with the machine had led to expensive warranty claims. Jowett decided to buy the company to recoup some of the money it was owed for engines. Bristol Tractors changed hands on 4 July 1935, with William Jowett becoming a director. Manufacture of the crawler moved to Blake Hill Works, at the rear of the Jowett factory in Idle, near Bradford. Slow

Roadless

RIGHT:
The Bristol crawler with the experimental tracks working with a set of discs.

BELOW:
All Ransomes MG crawlers were fitted with Roadless tracks. The model shown is an MG5 which dates from about 1949.

RIGHT:
The Roadless 25 – a prototype self-propelled toolbar for market garden use.

sales and reliability problems had resulted in only 270 machines being sold between April 1933 and June 1937 when an improved model was introduced.

After the war, the Bristol company was bought by the Austin vehicle distributors, H. A. Saunders, and a new model Bristol 20 was brought out with a four-cylinder Austin car engine. The Roadless rubber-jointed tracks were retained and remained a feature of Bristol crawlers until the Perkins-engined 25 came out in the late 1950s.

After being approached by Ransomes Sims and Jefferies of Ipswich to design them a garden cultivator, Roadless took the Bristol concept one stage further and drew up plans in December 1932 for an even smaller rubber-tracked crawler. Aimed at the market gardener, the project was known under the initials MG. The first MG crawler was pedestrian-controlled with the driver walking behind. This proved unsuccessful and was shelved. Work started on the MG2 in October 1933, and new designs were drawn up with the driver sitting on the machine, the advantage being that his weight gave the tracks added adhesion for no extra cost. A prototype was built and tested in 1934, and pre-production models underwent trials with Ransomes the following year. The MG2 was powered by a single-cylinder air-cooled Sturmey Archer engine developing 6 hp. This was a well-tried and tested power-unit, previously used by Ransomes on its larger lawn mowers. The transmission provided single forward and reverse gears. The little crawler was in production by 1936, and was demonstrated to the public for the first time on 29 April of that year. A total of over 15,000 MG2 and the subsequent MG5, MG6 and MG40 models were built, all using Roadless tracks.

Roadless also developed its own machine for market garden use. Designed as a small toolcarrier, it was powered by a single-cylinder air-cooled petrol engine. Known as the Roadless 25, it never went into production. Many of the tractor designs were attributable to Leonard Tripp, who had now taken over as chief designer, and their manufacture was supervised by Charles Skelton who had become works manager in 1934. Roadless had made its mark as a specialist tractor manufacturer, and was ready to expand further into the agricultural market as the clouds of war began to gather once again.

LEFT:
Charles William Skelton, works manager for Roadless from 1934 onwards.

BELOW:
Three of four Roadless tractors supplied to Earls Court for moving heavy exhibits. Pictured in 1937, are two Fordsons equipped with winches and a Bristol fitted with a front castor wheel to aid stability.

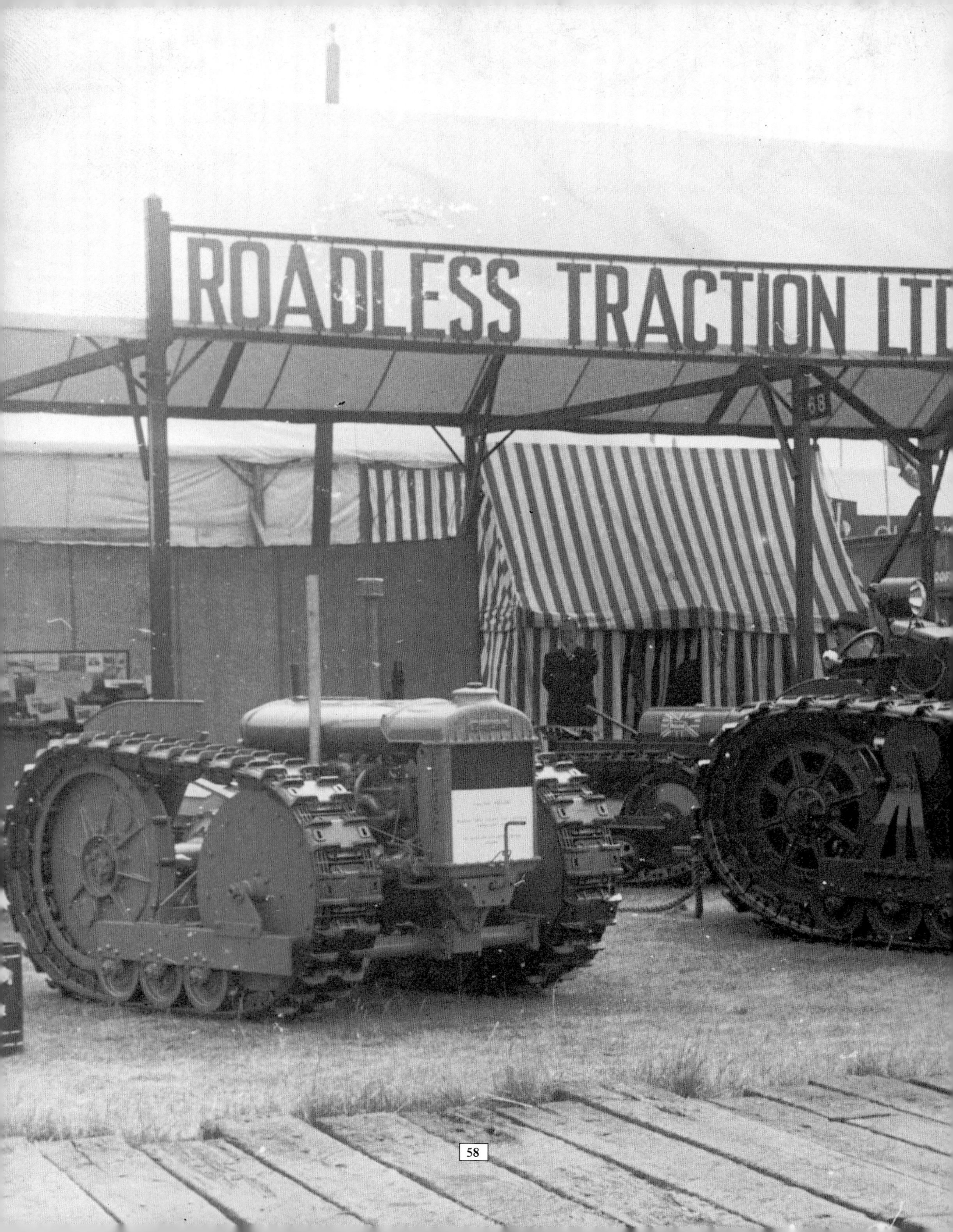
ROADLESS TRACTION LTD
68

CHAPTER 3
War and Peace
1939 to 1957

The Roadless stand at the 1939 Centenary Royal Show held in Windsor Great Park.

Roadless

RIGHT:
Roadless's secret weapon: a land torpedo developed in 1940.

BELOW:
The power unit for the land torpedo – an 8 hp side-valve Austin car engine.

The Second World War brought greatly increased production to Roadless. Roadless track-equipped tractors were extensively used by the Air Ministry for moving aircraft and maintaining runways on the many grass airfields that had been established across the country for the Battle of Britain and after, and Orolo track units were in great demand by different branches of the armed services for mounting and transporting various types of equipment – the searchlights used by Britain's anti-aircraft batteries had been mounted on Orolo track units since 1936 – while at the other end of the scale, heavy Orolo tracks were fitted to beach landing craft recovery cranes. Also, eighteen Case L tractors on Roadless tracks were fitted with Bray hydraulic angledozers and used by the Army in France and the Middle East.

By 1941, virtually all of the company's total production capacity was devoted to war work. Like many other factories, Roadless suffered air-raid damage. After one particularly devastating night bombing raid, the assembly and metal working shop collapsed and another part of the works was in such a dangerous state that it had to be pulled down. None of the buildings in the yard had escaped damage – all the windows and doors had been blown out, and the roofs needed repairing. Yet on the day after the raid, Roadless's deliveries were only down by

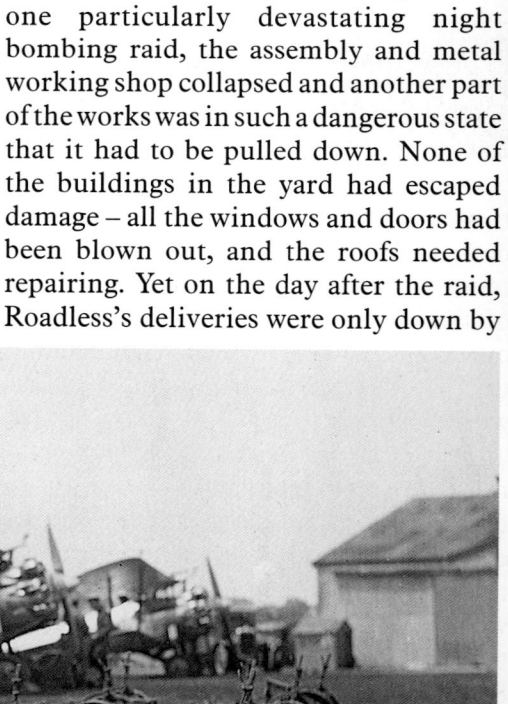

RIGHT:
A Rushton Roadless supplied to the Air Ministry in 1932 for mowing the grass at Biggin Hill aerodrome with a Ransomes gang-mower.

about 10 per cent. Such was the determination of the company not to be affected by the bombing that output actually rose during the days that followed as the employees, working without shelter and often in the pouring rain, battled to keep production going.

New buildings were erected at the first opportunity, but Roadless still marvelled at the devotion of its employees who always managed to turn up to work on time after many instances of them losing their homes and belongings overnight in the Blitz. *Roadless News* summed it up: 'The more they are bombed the harder they work.'

With Philip Johnson's background in military developments, it comes as no surprise to learn that Roadless was also involved in some experimental and highly confidential work for the War Office. Among the many facets of this work was the design and development of a land torpedo in 1940. This device was powered by an 8 hp side-valve Austin car engine suspended inside the body from a tubular backbone. Each end of the torpedo contra-rotated and, relying on the Archimedean screw principle, wormed itself across the ground towards its target. The nose-cone of the torpedo was packed with explosive designed to detonate on impact. However, field trials showed that when crossing ditches or trenches, the device had a worrying tendency to turn over and head back to where it had come from, and so the project was dropped.

Roadless at war will be best remembered for the many tractors on half-tracks that were to be

LEFT:
One of several Fordson Roadless tractors ordered by the Air Ministry during 1936.

BELOW:
1939 Fordson Roadless half-track fitted with a Hesford Minor winch.

Roadless

RIGHT:
One of many Fordson Roadless half-tracks fitted with front-mounted Hesford winches that were used by the RAF during the Second World War

RIGHT:
A wartime Fordson Roadless half-track Note the forecarriage arrangement.

seen hard at work on Britain's airfields. Roadless supplied its first tractor, a Rushton Roadless, to the Air Ministry as far back as 1932 for mowing the grass at Biggin Hill aerodrome with a Ransomes gang-mower. Three more Roadless equipped tractors were bought by the Ministry the following year. Fears of the gathering might of the German war machine led to plans being announced in May 1935 to treble the size of the Royal Air Force over the next two years – an achievement which was accomplished by January 1937. This expansion led to a large order for Roadless tractors being placed by the Air Ministry during 1936 – the year the Spitfire came out. This new order was for Fordson Roadless tractors fitted with electric lighting to

War and Peace

LEFT:
A Fordson Roadless half-track that saw service with the United States Air Force.

BELOW:
Fordson Roadless half-track fitted with a Hesford front-mounted winch. The chain drive arrangement can be clearly seen.

enable them to work through the night. The tractors were supplied through the London dealers, W. J. Reynolds of East Ham, and were distributed to various airfields by April of that year.

The RAF encountered some problems with the tractors pitching when mowing at speed. To overcome this, in 1938 Roadless introduced a forecarriage arrangement with a front axle and wheels which turned the crawler into a half-track. The standard rubber-jointed tracks were still used, but the front wheels, which now controlled the steering, made the tractor more stable and easier to drive. Another advantage of the forecarriage arrangement was that it provided a platform for front-mounting attachments such as winches and cranes.

The Fordson Roadless half-tracks were supplied in considerable numbers to the Air Ministry, and provided sterling service with the RAF throughout the war years. The tractor was used for far more than just mowing the airfields and was employed moving aircraft and supplies, hauling fuel bowsers and bomb trailers, and even sweeping the runways clear of snow. Many of the tractors supplied to the RAF were fitted with

LEFT:
Case DEX Roadles half-track fitted with a Morris 1-ton jib crane in 1941. The crane cost £47 10s 0d extra.

63

Hesford winches for recovery work, and others had front-mounted jib cranes for dealing with crashes or changing aircraft engines. Twenty of the half-tracks were also supplied for Operation Dynamo and were used on the Channel beaches hauling boats and vehicles out of the water following the evacuation of Dunkirk in June 1940.

In July 1941, Roadless announced that the half-track arrangement was now available on the Case DEX tractor, coming into the country following the Lend-Lease Act passed by the United States Congress in March which allowed Britain to draw on the resources of its

ABOVE:
Case DEX fitted with Roadless full-tracks.

RIGHT:
Fordson N fitted with DG4 half-tracks. The background to this 1944 shot gives an interesting view of the Roadless yard at Hounslow.

BELOW:
A 1945 Fordson N with the benefit of Roadless DG4 half-tracks and a driver's canopy which could also be bought from the company for £11 10s 0d.

transatlantic ally. The 30 hp DEX model was specially designed to suit the particular needs of Britain's farmers, and was not marketed outside the United Kingdom. The Roadless half-track version proved as versatile as the Fordson and several were supplied to the Air Ministry, including nine with Morris 1-ton jib cranes. At least one Case DEX was also built as a full-track.

Although Roadless advertised the fact that the Fordson and Case half-tracks were suitable for agricultural and logging work, all that Hounslow could produce were swallowed up by the Ministry and none were available for civilian use during the war. However, after the war finished, many of these machines became surplus and

ABOVE:
This Fordson N on half-tracks has been fitted with a telescopic front axle to allow it to plough 'on the land'. Roadless works manager, Charles Skelton, at the wheel of the tractor, is watched by farmer and inventor, Mr A. J. Hosier, and Lieutenant-Colonel Philip Johnson, who is on the extreme right of the photograph.

LEFT:
Case C fitted with Roadless DG4 half-tracks in 1945.

RIGHT:
DG4 tracks fitted to a Case DEX tractor.

found their way on to farms and in to forests. As the war drew to a close, and the pressure was taken off the company, Roadless managed to produce several more Fordson full-tracks for agricultural use.

Roadless's conception of its products had changed somewhat since the early days of the company. In 1942, India had been promised postwar independence and the British Empire was beginning to dismantle. War had altered the world, and the vision of Britain needing fleets of off-road vehicles to expand its industries into the vast continents of the world was fast becoming irrelevant. But the world needed feeding, and so

RIGHT:
The Case LA came out on the larger Roadless DG8 tracks in 1946, but the tractor's bottom gear was found to be too fast to use them to their best advantage.

did Britain. Postwar agriculture had to catch up with wartime shortages. Increased production was required from farms faced with limited manpower. The tractor would be more important than ever – it needed to be more efficient and capable of bringing the world's undeveloped land into production. This was the challenge Roadless believed it now faced.

Roadless always felt that the half-track's performance could be greatly improved, and by 1943 was looking into ways of adopting the 'big wheel' principle of the locked girder track, used so successfully in the Orolo units, to suit a tractor. The result was the Driven Girder, or DG, half-track which was designed to lock in a predetermined curve equal to a wheel of 20 ft diameter.

ABOVE:
An American Oliver 80 Standard tractor fitted with Roadless DG4A half-tracks in 1946.

LEFT:
Roadless DG4A half-tracks fitted to a Canadian built Massey Harris 102 Junior Twin-Power tractor.

BELOW:
This Allis Chalmers Model U was converted to DG4A half-tracks by Roadless for a customer in 1946.

The new half-track first appeared in 1944 in DG4 form on the Fordson tractor. The DG4 tracks performed well; the tractor's drawbar pull was greatly increased, and it was easier to steer than the old half-track system. Roadless also offered a telescopic front axle, suggested by Mr. A. J. Hosier and built by Stanhay of Ashford, Kent, which allowed the half-track to plough on the land with one front wheel in the furrow.

Another advantage of the DG halftracks was that they could easily be fitted to an ordinary wheeled tractor without any modification. The drive sprockets were fitted in place of the rear wheels, and the axle carrying the two idlers was mounted on a bracket bolted to the rear transmission housing. The instruction book for the DG tracks stated: 'any competent mechanic with the assistance of a helper can mount the DG equipment on to a wheeled tractor in from two to four hours' time, without any equipment other than the jacks and hand tools always available at a garage and very often on the farm.' It was not always that easy in a mud-soaked farmyard in a freezing January gale.

In 1944, DG6 tracks were built for the FWD lifeboat tractors that were still working, and a set of the same tracks were modified to fit an old McLaren motor-windlass used for cable ploughing. The

ABOVE:
1947 David Brown Cropmaster fitted with Roadless DG4 half-tracks

RIGHT:
A Fordson E27N Major on Roadless DG4 half-tracks clearing scrub with heavy discs in the Ohariu Valley, near Wellington, New Zealand, in 1949. The New Zealand Minister of Agriculture, Mr Cullen, is at the wheel.

BELOW:
Fordson E27N Major on Roadless half-tracks hauls 12 tons of logs at a Norwegian lumber camp in the Glommen Valley in 1949.

same year, Roadless produced a diminutive set of half-tracks for the Trusty garden tractor, built by Tractors (London) Limited, agents for the former Rushton tractor. In July 1945, the Case DEX was also offered with DG4 half-tracks, and over the next couple of years similar tracks were fitted to Case C, Massey Harris 102 Junior, Oliver 80 and David Brown Cropmaster tractors. Case L and LA tractors were put on the larger DG8 tracks, and one Allis Chalmers Model U was fitted with DG7 tracks for a customer. Drawings were made of DG4 tracks for the John Deere D and DG8 tracks for the Oliver 90, but no photographs exist of these machines, and it is unlikely any were built.

It was the Roadless conversion of the Fordson E27N Major that really put the half-track on the map. This new model from Ford rolled off the Dagenham assembly lines in March 1945, and Roadless had DG4 track equipment for it ready by September. The price of the equipment to fit either the Fordson N or E27N was set at £175.

The Roadless conversion was sold as a Ford-approved attachment and was advertised in many of Ford's publications. The E27N proved to be a very successful model for Ford and nearly a quarter of a million were built. It was exported all over the world – some 30,000 going to Australia alone – and nearly everywhere the Fordson was sold, the Roadless half-track version was available, proving its worth on all sorts of terrain from the Australian outback to forestry plantations in Norway. Roadless was further encouraged when its half-track equipment for the E27N was awarded a silver medal by the RASE at the 1948 Royal Show at York.

The introduction of the Perkins P6 diesel engine option for the E27N in 1948, with its increased power and pressurised lubrication system, made sales of the Roadless half-track even more popular. The tractor could now work on slopes that would have caused oil starvation problems for the old TVO engine relying on gravity-fed lubrication.

A new skeleton half-track, with the standard track plates replaced by spuds, was introduced

TOP:
Fordson E27N Major on Roadless skeleton tracks.

ABOVE:
A comparison in width between the standard Fordson E27N half-track and the narrow model introduced by Roadless for orchard work in 1950.

LEFT:
A line of Fordson E27N Major Roadless half-tracks fitted with Perkins P6 diesel engines await export in the yard at Hounslow.

RIGHT:
Fordson E27N Major fitted with Roadless DG4 half-tracks.

BELOW:
Sometimes even a Roadless half-track needs a helping hand. A Field Marshall Series lll assists a Fordson E27N on DG4 tracks with a set of heavy Ransomes discs at Manor Farm, Denton, Norfolk, in 1966.

LEFT:
An illustration of the Fordson E27N Major on half-tracks taken from the 1946 Roadless calendar.

BELOW:
The Roadless works in the garden of Gunnersbury House, Hounslow, photographed in 1947. In the foreground is the recently erected flame cutting, welding and assembly shop.

BOTTOM:
Fordson E27N Major half-tracks under construction in the Roadless assembly shop in 1947.

in early 1949 for seedbed cultivations and working on certain types of soil. The advantage of this track was that the special spud plates lifted and gently cultivated the soil as the tractor moved, virtually eliminating wheelings. A narrow version of the Fordson E27N Roadless half track was released in 1950. This machine was jointly evolved by Roadless, the Ford Motor Company and Ford Afrique of Algiers, and was designed for working in vineyards. Both the track equipment, and the rear axle housings, half shafts and front axle were modified to give an overall width of only 65 in.

Although the company was experiencing considerable success with the DG tracks for the Fordson E27N, Roadless half-tracks were still made for a number of other machines. An experimental Field Marshall Series ll tractor fitted with DG4 tracks underwent trials at Gainsborough in 1948, and evidently performed beyond all expectations. At least one other was built on skeleton tracks, but adverse feeling from the Fowler side of the Marshall company, who felt it would take sales away from their full-track VF model, led to it being dropped.

The DG13 half-tracks were designed for smaller tractors including the Allis Chalmers Model B, the Newman built at

Roadless

*RIGHT:
An experimental Field Marshall Series II tractor fitted with Roadless DG4 half-tracks.*

*BELOW:
Field Marshall Series II on Roadless skeleton tracks.*

Grantham, and the Ford Ferguson, 8N and NAA tractors. One Ford Ferguson 2N was built up on these tracks in 1944, and a number of Allis Bs were converted after the equipment was marketed through Allis Chalmers agents, priced at £218, from 1952. Roadless also experimented with a low orchard version of the Allis B on half-tracks.

Apart from the Ford conversions, Roadless enjoyed most success with DG4A half-track equipment for the British-built Massey Harris 744D tractor, which was also powered by the Perkins P6 engine. The conversion, available from 1951, cost £247 with the standard track, or £258 for the skeleton. A considerable number of the conversions were sold, and one farm in Hampshire ran four Massey Harris 744Ds on Roadless half-tracks.

From 1951, several DG4 half-track conversions were sent to Italy for use on the Landini L45, a single-cylinder semi-diesel tractor of 45/50 hp. DG13 tracks were supplied the following year for the smaller L25 tractor, and the more powerful L55 model eventually gained the benefit of DG15 tracks. DG track equipment was also designed for the Austrian Steyr-Daimler-Puch and Australian Chamberlain tractors in the early 1950s, and at least one Lanz Bulldog came out on Roadless half-tracks. Roadless continued to supply a few track sets to

*ABOVE:
Roadless DG13 half-tracks fitted to a Ford Ferguson 2N in 1944.*

*RIGHT:
The Allis Chalmers Model B was available on Roadless DG13 half-tracks from 1952.*

Tractors (London) Limited for its improved Trusty Steed model.

The new Fordson E1A Major was released in November 1951. This much improved tractor now had the advantage of an excellent overhead valve diesel engine that was to prove a dependable and tireless power unit. Roadless, naturally keen to promote its half-track equipment for the new tractor, had an E1A Major on DG4 tracks ready for the December Smithfield Show. The company

ABOVE:
A prototype orchard model Allis Chalmers Model B on experimental Roadless tracks in 1946.

LEFT:
This hybrid machine on Roadless skeleton tracks appears to be a Canadian Massey Harris 44 fitted with a Perkins P6 diesel engine. The photograph pre-dates the British-built 744D tractor and is probably a prototype.

LEFT:
A Massey Harris 744D on Roadless half-tracks struggles to cope with some appalling wet conditions.

RIGHT:
A Massey Harris 744D fitted with Roadless skeleton tracks.

RIGHT:
An Italian Landini L45 single-cylinder semi-diesel tractor on Roadless DG4 half-tracks.

BELOW:
Roadless skeleton tracks help this Landini L45 cope with land that has turned to mud on this Italian farm.

had been given plenty of prior notice for the release of the new tractor, and had in fact been working on half-track arrangements for the prototype since 1949.

The DG4 tracks were superseded by the improved DG15 tracks in 1953. The DG15 equipment for the Fordson Major cost £270, and the complete Fordson Diesel Major on half-tracks was advertised by Roadless as 'The Cheapest 40 hp Crawler in the World' at £766 5s 0d. The new Major was also available with skeleton, or composite semi-skeleton, tracks. To enable the half-track to be used with mounted implements, Roadless designed a lift linkage which made it possible for the Major's hydraulic power-lift to still be used when the tractor was on DG tracks.

The Fordson Major on DG15 tracks became the basis of a couple of conversions made by other manufacturers. The Gibbon Roadless lime spreader was made by Atkinson's Agricultural Appliances Limited. First seen at the 1954 Smithfield Show, it consisted of a Fordson on half-tracks fitted with a 2-ton Atkinson hopper. The Bray Half-Track Hydraloader, built by W. E. Bray of Feltham, came out in 1955 and was also based on a Fordson on DG tracks.

After an agreement was reached with Morris Motors Limited, the DG15 tracks became available in 1957 for fitting to the Nuffield Universal tractor, but not very many were sold. This appears to have been the last new application for the DG tracks, as the company then began to concentrate on four-wheel drive, although the Fordson half-track was to remain in production for a few years to come.

Roadless had not completely forsaken the rubber-jointed track during this time. It had been steadily improved and was still supplied to various manufacturers, including Bristol and Ransomes. Fraser Tractors of Acton also used Roadless tracks on its small crawler brought out in May 1950, but these were of the rigid-girder design. Also, in 1946, a new model of Roadless Lifeboat tractor was introduced for the RNLI, based on the Case LA skid-unit.

Roadless could still see a future for full-track crawlers on rubber-jointed tracks. In 1947, the company had started work on a full-track version of the Fordson E27N Major, using a modified

ABOVE:
A German Lanz Bulldog 45 hp tractor fitted with Roadless half-tracks in the yard at Hounslow.

LEFT:
Also seen in the yard at Hounslow is this Austrian Steyr tractor on Roadless half-tracks.

BELOW:
The new Fordson E1A Major fitted with Roadless DG4 half-tracks in 1951.

Roadless

RIGHT:
A 1953 Fordson Diesel Major on Roadless skeleton tracks ploughing on the Hoggsback near Guildford.

BELOW:
A 1953 Nuffield Universal DM4 fitted with Roadless DG15 half-tracks.

RIGHT:
An improved version of the Roadless Lifeboat tractor was introduced in 1946, based on the Case LA skid-unit. This example, KLA 84, is seen in service with the RNLI at Newbiggin Lifeboat Station, Northumberland, in 1950.

version of the E3 rubber-jointed tracks. News of an intended County crawler on the same skid-unit must have reached Roadless and possibly prompted its actions.

The prototype machine, fitted with E3B tracks with three bottom-rollers, underwent trials with the Ford Motor Company early in 1948. Unlike the earlier Roadless crawlers, it featured clutch and brake steering. Roadless, conscious of the pitching problems encountered with the earlier Fordson crawlers, had tried to keep the weight of the machine forward by placing a spacer-box between the engine and transmission. Unfortunately, the tractor was still found to be unbalanced, so in July Roadless returned to the drawing board and drew up plans for a revised model with four-bottom rollers to

LEFT:
Roadless Lifeboat tractor, KLA 84, brought back from Ireland in 1996, is currently under restoration in the hands of the present Newbiggin Lifeboat mechanic and tractor driver, Richard Martin.

LEFT:
One of the production Roadless Full Track Model E crawlers at work in Dorset.

BELOW:
A derelict, but rare, Roadless Full Track Model E awaits restoration in Lincolnshire.

aid stability.

After exhaustive testing, details of the new tractor, known as the Roadless Full Track Model E, were released in July 1950. The company announced that the machine would be available with either the Fordson TVO or the Perkins P6 diesel engine. Prices were provisionally set at £850 for the TVO version, and £1,150 for the Perkins-powered tractor.

Roadless's sales manager, Captain K. Knights, stated that the company's intention was only to build the Model E in limited numbers for the home market, and none would be available for export. In the event, the new crawler was produced in very limited numbers, with estimated figures as low as twenty-five sometimes being given. The reason for this low

Roadless

*RIGHT:
A prototype Roadless Full Track Model E based on the Fordson E27N Major skid-unit.*

*BELOW:
The prototype Roadless Full Track Model E at Boreham House for trials with the Ford Motor Company.*

output is not known. Maybe the company had encountered more problems with the machine, or felt the slightly cheaper County Full Track had already cornered the market, or more likely, Roadless were at full stretch to cope with orders for the DG half-tracks.

Following the introduction of the E1A Fordson Major in 1951, Roadless started work on a new full-track crawler version of this new model. The new crawler was seen for the first time at the 1953 Smithfield Show in December. Designated the J17, it incorporated a new type of rubber-jointed track using a scaled-up version of the track used for many years on the Ransomes MG crawlers. Pre-production trials were held on farms in East Anglia, and Metheringham near Lincoln. The tractor went on general sale in August 1954, priced at £1,295.

The J17 enjoyed more success than its predecessor, the Model E. It proved popular on the fenland farms, and by the end of 1955, nearly 100 were working in East Anglia and Lincolnshire. Production of the J17 carried on into the early 1960s and was based on the subsequent new Fordson Power Major and Super Major skid-units, but in dwindling numbers as four-wheel drive sales took over.

Roadless also built a smaller prototype R20 crawler for the South American market. The original experiments for the R20 were carried out in 1950 on an American Cleveland Cletrac HG fitted with rubber-jointed tracks. The fully developed prototype appeared in August 1954 at the Royal Show held at Windsor. Fitted with J13 tracks, it had a

LEFT:
The prototype Roadless Full Track Model E. Note this tractor has only three bottom track-rollers unlike the production model which had four.

LEFT:
The prototype Roadless Model E crawler at Silsoe for testing by the NIAE.

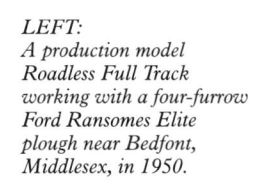

LEFT:
A production model Roadless Full Track working with a four-furrow Ford Ransomes Elite plough near Bedfont, Middlesex, in 1950.

ABOVE:
Roadless Traction's stand at the 1950 Royal Show. Nearest to the camera is the Full Track Model E. The large diameter wheel in the background is to illustrate the size of wheel that would be required to duplicate the performance of the Roadless half-track.

RIGHT:
The Roadless J17 crawler, based on the Fordson Diesel Major and introduced in 1953.

RIGHT:
The prototype Roadless J17 on trial. Charles Skelton is leaning on the back of the tractor and Philip Johnson can just be seen in the centre of the group.

6-speed gearbox and was powered by a Perkins P3 diesel engine.

During the early 1950s, Roadless managed to find time to develop a number of other products apart from the track systems, including a high-clearance front axle arrangement with large diameter tyres for forestry applications. This conversion started life on a Fordson E27N which was developed in 1951 for travelling through swampland hauling spraying equipment for locust control in Tanganyika. This machine was fitted with a special cranked front axle which allowed wheels with 11 in. x 36 in. tyres to be fitted all round for increased rolling resistance and mobility on soft terrain. It was used with a trailer made by Taskers of Andover that was equipped with the same tyres.

In 1952, Roadless carried out trials in conjunction with Colonel R. G. Shaw, the machinery research officer for the Forestry Commission. These trials, conducted at the Commission's experimental station at Alice Holt Forest in Hampshire, involved a Fordson E27N on DG4 tracks fitted with a similar

War and Peace

LEFT:
The Roadless J17 at work near Metheringham in Lincolnshire with a Ransomes 13-tine Dauntless cultivator in 1954.

LEFT:
Roadless J17 crawler.

81

RIGHT:
A Roadless J17 with a set of gang rolls at Woodton Farm near Norwich in 1964. The cab had come from a scrapped Fowler VF crawler.

RIGHT:
An American Cleveland Cletrac HG used as a test-bed for track designs for the Roadless R20 project.

War and Peace

front axle on 11 in. x 36 in. tyres. Pulling a four-wheel timber trailer, also fitted with oversize tyres, the unit managed to cope with negotiating water, slush, ditches and fallen logs without any problem. Roadless later marketed the conversion to suit the E1A Major from 1953, using a slightly different axle arrangement with smaller 10 in. x 28 in. tyres. The complete conversion cost £166.

Other products from the company during this period, included a portable dynamometer and the Servis-Roadless brush cutter. This American-designed heavy-duty rotary brush cutter was made under licence from the Servis Equipment Company of Texas. It also underwent

LEFT:
The prototype Roadless R20 crawler. Fitted with a Perkins P3 engine, this tractor was brought out in 1954 for the South American market.

LEFT:
Developed in 1951 for locust control in Tanganyika, this special high-clearance Fordson E27N and Taskers trailer were fitted with 11in. x 36in. tyres all round for greater mobility over swampland.

*RIGHT:
Fordson E27N Roadless half-track fitted with a high-clearance front axle and 11 in. x 36 in. tyres for use by the Forestry Commission.*

*RIGHT:
The Roadless half-track with the high clearance front axle undergoing trials with the Forestry Commission in Alice Holt Forest with a specially converted Taskers log trailer.*

*RIGHT:
Fordson E1A Major on half-tracks, fitted with the Roadless high-clearance front axle conversion with 10 in. x 28 in. tyres that was introduced in 1953.*

the USA, for inter-row cultivation and maize harvesting, and the sugar cane and cotton growing areas of other countries of the world.

To rectify this, Ford approached Roadless to build them a tricycle conversion for the Major. Roadless started work on the project in February 1952. After much collaboration with the design and engineering departments at Ford Motor Company, a trials with the Forestry Commission for clearing undergrowth.

Soon after the launch of the new E1A Major, the Ford Motor Company in Britain had reached an agreement with the Tractor and Implement Division of the parent company in the USA for the Diesel Major to be sold on the American market. Unfortunately the new Fordson Major was at a disadvantage; like its predecessor, the E27N, it was not available from the factory in the tricycle rowcrop configuration demanded in certain parts of

LEFT:
The prototype Roadless tricycle rowcrop conversion fitted to a Fordson E27N Major.

first prototype was completed on an E27N skid-unit. The prototype had a tricycle front axle and extendable rear half-shafts to give adjustable wheel tread. The front axle arrangement was a little cumbersome and increased the wheelbase too much. In the meantime, the first consignment of Fordson Diesel Majors (in standard configuration) had gone to the USA in September 1953.

After more design work, a Diesel Major with an improved front axle pedestal was shown on the Ford Motor Company stand at the 1954 Smithfield Show. The first consignment of fifty rowcrop tractors was completed in August 1955. Forty-nine went to the USA and one was sold in Britain. In September 1955, Roadless released details of the final design with an extendable rear axle that would adjust to 96 in. tread.

LEFT:
The extendable rear axles are clearly demonstrated in this photograph of the prototype rowcrop tractor.

ABOVE:
The final design for the Roadless rowcrop conversion fitted to a Fordson Diesel Major

RIGHT:
Another view of the Fordson with the Roadless tricycle conversion. This tractor was exhibited at the 1954 Smithfield Show.

The conversion was supplied for many of the Fordson tractors sold in the USA up to 1964. At $2,700, the rowcrop Fordson compared very favourably with other diesel powered tractors in the USA. The Case 500 cost over $5,000 and was not available in rowcrop configuration.

The first Roadless four-wheel drive tractor came out in 1956. The long years of track production were drawing to a close at Hounslow and four-wheel drive developments began to dominate the company's work. Ironically, it is this latter work for which the company is best remembered, yet it produced vehicles on tracks for over four decades.

ABOVE:
The Servis-Roadless heavy-duty brush cutter.

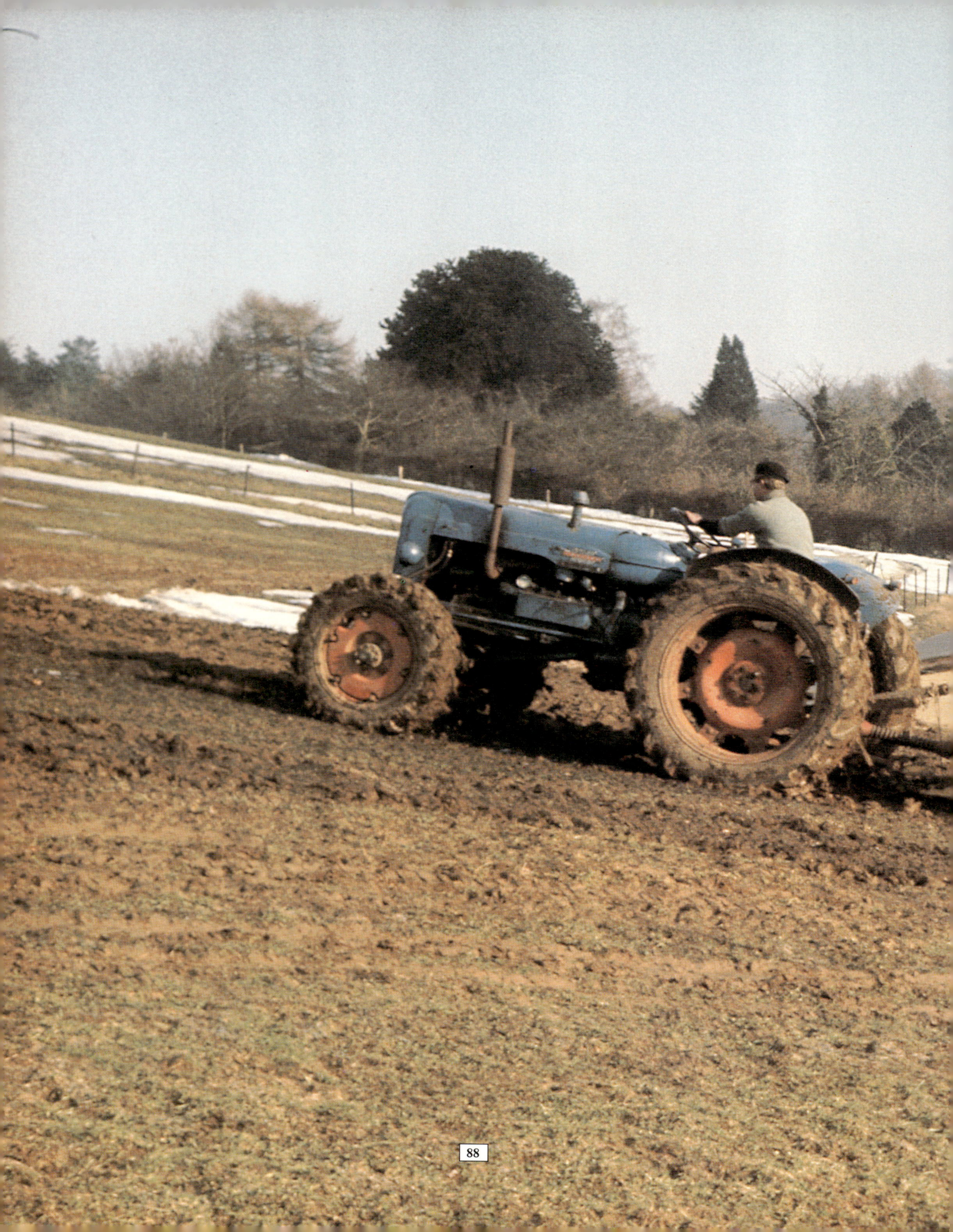

CHAPTER 4
Four-wheel Drive Developments 1956 to1964

A 1960 Fordson Power Major with the Roadless four-wheel drive conversion spreads chicken manure in Surrey.

*RIGHT:
During one of his many overseas trips, Philip Johnson visits the sisal growing area of Tanganyika to see a Roadless half-track at work.*

*BELOW:
A Fordson Diesel Major fitted with one of the first Roadless four-wheel drive conversions.*

Philip Johnson, now both chairman and managing director of Roadless, enjoyed travel, and felt the need to visit most of the areas of the world where the company's tractors were at work. Through these trips, he gained an important insight into the local conditions and problems under which the machines would operate. It was during one such visit that a chance meeting led to Roadless's long involvement with four-wheel drive tractors.

During 1952, Johnson travelled through Italy to see the Landini tractors on DG half-tracks working – a trip which included a five-hour train journey from the north of the country, through the Appennines to Rome, before visiting the Landini factory at Fabbrico. Landini were so

Four-wheel Drive Developments

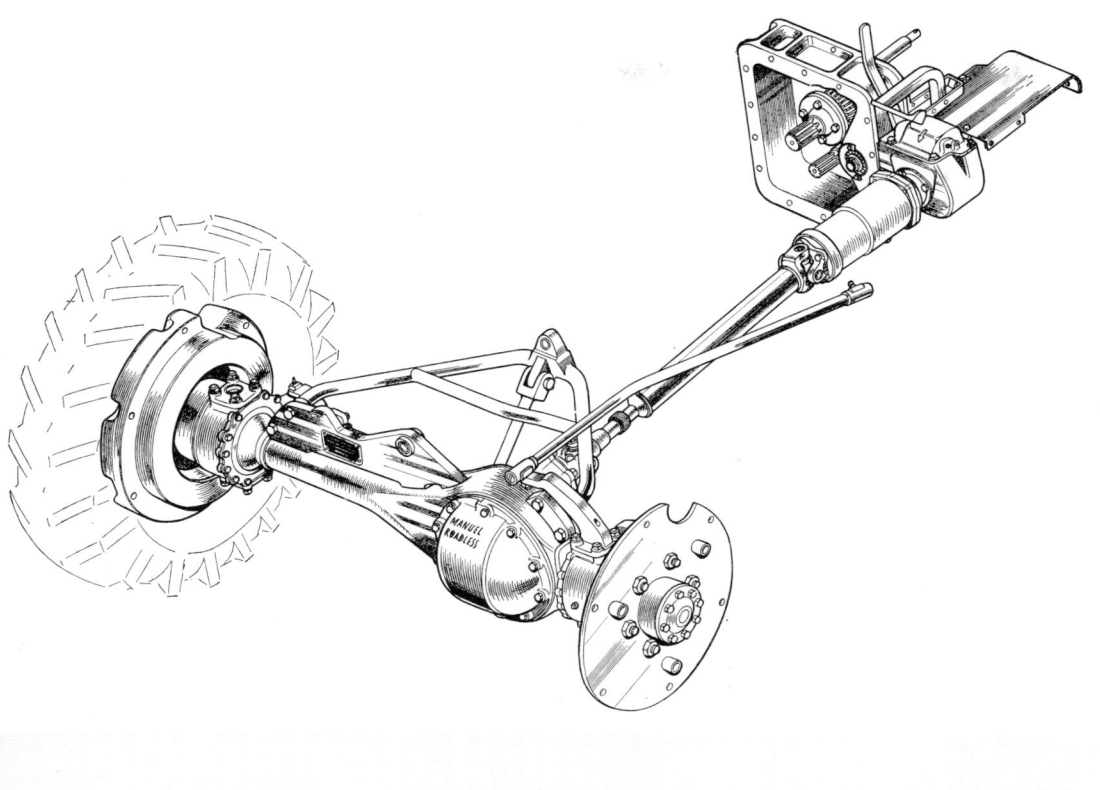

*LEFT:
Exploded drawing showing the Roadless transfer box assembly and four-wheel drive front axle for the Fordson Major.*

*LEFT:
The Roadless four-wheel drive demonstration tractor based on a Fordson Power Major.*

pleased with the performance of the DG tracks, that Johnson was presented with a photograph album of the tractors at work.

While in Italy, Johnson met with Dr Segre-Amar, founder of the Selene S.A.S. company of Nichelino near Turin. Selene had pioneered the use of four-wheel drive on Fordson tractors by developing a conversion using a transfer box

ABOVE:
An ex-US Army GMC 2½-ton 6 x 6 truck of the type used by Roadless as a source for its four-wheel drive front axles.

RIGHT:
A close-up of the Manuel Roadless front axle.

sandwiched between the gearbox and rear transmission, driving an ex-US Army war-surplus four-wheel drive front axle. The original conversion was evolved by Selene to boost the sales of some old model Fordson E27N tractors the company had on its hands when the new E1A Major came out in 1951.

Segre-Amar and Johnson had mutual interests in the agricultural engineering field, and soon built up a rapport that was to extend into a long friendship. Towards the end of 1952, Segre-Amar paid a visit to England for the Smithfield Show. While on the Ford Tractor Division's stand, he was introduced to William Batty (later Sir William Batty and UK Chairman of the Ford Motor Company) by a mutual friend, Mr S. B. Handley of the Northamptonshire Ford tractor dealers, E. Ward (Wellingborough) Limited. Segre-Amar and Batty discussed the possibilities of the Selene conversion being fitted to Fordson tractors in Britain. After hearing Segre-Amar out, Batty is quoted as having said, 'Go ahead. I don't promise to help, but certainly I won't interfere with your plans.' Furthermore, while not wishing to become involved in the conversion, Ford agreed to maintain the guarantee on any tractors fitted with the four-wheel drive system.

Encouraged by Ford's attitude, Segre-Amar then established patents for the system in Italy and contacted Johnson with a view to Roadless producing the Selene conversion in the UK. At Johnson's suggestion, a Fordson Major fitted with Selene four-wheel drive was sent from Italy to the National Institute of Agricultural Engineering at Silsoe for testing and evaluation in 1954. The NIAE test report, published in December, showed very favourable results and highlighted the tractor's excellent performance.

In the same month, Selene applied to the London Patent Office for the grant of a British patent to cover its four-wheel drive conversion for the Fordson and other tractors. The application for 'Improvements in or relating to Transmission Gearing for Tractors' was filed under no. 770,992. The patent only covered the transfer box and the drive to the front axle, as the axle itself was a war-surplus part and not of Selene design.

During 1955, Philip Johnson went back to Italy to stay with Segre-Amar, and on his return, after further correspondence, an agreement was reached for Roadless to manufacture the Selene conversion under licence in the UK. By this time, Selene had designed four-wheel drive systems for other makes of tractor, but Roadless only gained the rights to build the Fordson conversion. An arrangement was made for the licence to be extended to other makes of tractor from time to time, but only using the drive system as covered by patent no. 770,992.

Roadless's licence gave it the exclusive rights to sell the four-wheel drive equipment in the UK

ABOVE:
A 1956 Fordson Diesel Major with Roadless four-wheel drive.

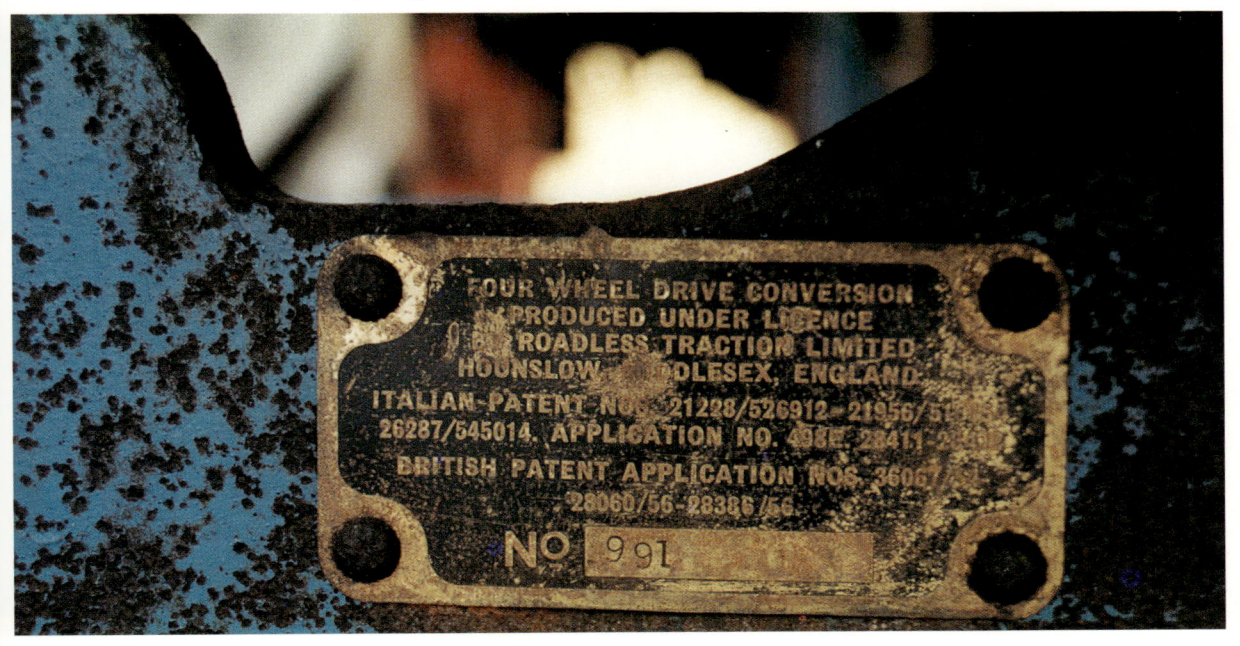

LEFT:
The manufacturer's plate fixed to the Roadless four-wheel drive axle giving details of the patents under which the conversion was made.

93

RIGHT:
A Roadless four-wheel drive Power Major that was sent to the NIAE at Silsoe for testing in 1959.

BELOW:
A Fordson Power Major with Roadless four-wheel drive hitched to a Bomford Superflow heavy cultivator.

and British Commonwealth, with the exception of New Zealand and India. Selene reserved the rights to sell its products exclusively in Italy, France, Spain and their respective dependencies. Roadless and Selene were then free to compete with each other in the remaining countries of the world, although Roadless was bound by an agreement to keep its net prices no less than 20 per cent higher than Selene's in t he areas of competition.

Under the terms of the licence, Roadless paid Selene a royalty of £20 for every conversion built at Hounslow. The figure of £20, consisting of £14 for patent use and £6 for technical assistance, was roughly an agreed 5 per cent of the £460 selling price of the conversion. A provision was made for a small payment to go to Mr S. B. Handley at E. Ward (Wellingborough) Limited for his part in establishing the Selene system in the UK. Handley also had the final decision in any arbitration between the two parties. It is a good indication of the mutual trust that existed between Selene and Roadless that the formal agreement covering the use of the licence

LEFT:
A 1957 Roadless four-wheel drive Major fitted with a Leeford hydraulic Muledozer.

LEFT:
A Roadless four-wheel drive Power Major fitted with a Sta-Dri cab made by Bristol Metal Components Limited.

ABOVE:
A 1963 Roadless four-wheel drive Super Major fitted with a Whitlock Dozaloda. This tractor has the industrial front axle with planetary hubs.

RIGHT:
A muckspreader towed by a 1960 Roadless 4WD Power Major is loaded with chicken manure by a Roadless Super Dexta in Surrey.

was not drawn up until 24 September 1959 – over three and a half years after Roadless had built their first four-wheel drive tractor using the Selene system.

As a point of interest, correspondence between Roadless and Selene shows that Roadless's competitors, County, were interested in adopting another different Selene four-wheel drive system in 1961, but Selene assured Roadless that the Fleet company would be paying very much higher royalties. In the event, County developed its own four-wheel drive system. Roadless exhibited one of Selene's four-wheel drive Fordsons at the 1955 Smithfield Show, and released details of the intended production of the tractor, which was to be known as the Manuel-Roadless. Manuel was Selene's trade name for the conversion, and was stamped on the axle casing of the early Roadless four-wheel drive tractors.

The Manuel-Roadless Fordson Major went into production the following year, and could be supplied as a complete tractor or as a conversion kit. The price for the complete tractor in diesel form was set at £978. The first tractor was completed on 6 March 1956, and supplied through the Ford Dealers, J. E. Coulter Limited of Evesham. Although the design was Selene's, all the parts for the conversion were manufactured at Hounslow. The design was simple; spur gears in the transfer box, which was mounted between the gearbox and the rear transmission and driven off the gearbox output-shaft, transmitted the drive to the front axle via a Hardy-Spicer propellor-shaft. The propellor-shaft was driven through a torque-limiting clutch to safeguard the front axle from overloading. A hand lever on the left of the transfer box operated on a sliding gear mechanism to engage or disengage front-wheel drive. The advantage of the system was that the gear ratios in the transfer box could be easily altered to suit different wheel equipment.

Like Selene, Roadless used ex-US military vehicle front axles modified to suit the tractors. The axles used were the banjo-type as fitted to the GMC 2½-ton 6 x 6 truck, produced in 1,000s for the US Army during the Second World War. The original agreement was for Roadless to buy the axles through Selene, but it was unlikely that the Hounslow company stuck to the agreement for any length of time, as supplies of the axles, new or reconditioned, were widely available from the many war-surplus

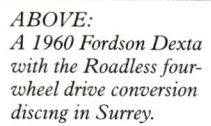

ABOVE:
A 1960 Fordson Dexta with the Roadless four-wheel drive conversion discing in Surrey.

LEFT:
Roadless Traction Limited's stand at the 1959 Smithfield Show. Centre left is the Roadless Angledozer based on a four-wheel drive Power Major fitted with Bray equipment. The Roadless Dexta is on the right of this machine.

RIGHT:
A Roadless four-wheel drive Super Major fitted with a Sydelift bale loader is seen on the company's stand at the 1960 Smithfield Show. In the background is a Roadless Dexta.

RIGHT:
Roadless four-wheel drive Super Dexta fitted with a Horndraulic loader.

dumps and ex-military equipment dealers in Holland, Belgium and across the rest of Europe, for as little as £7 per unit.

The Manuel-Roadless four-wheel drive tractor met with instant acclaim. Only about thirty tractors and kits were sold in the first year, but in 1957 well over 100 were dispatched. It was quickly adopted for agricultural, industrial and logging applications. Many were fitted with front-end loaders, angledozers, or winches for timber extraction. Roadless developed a heavy-duty front axle for use with loaders and other similar front-mounted attachments. Unlike the standard axle, which was shortened from the original GMC by cutting and welding, the heavy-duty axle would take a full-length half-shaft and became known as the 'wide axle'. This axle was used on a four-wheel drive loading shovel, based on the Power Major and built by Roadless in conjunction with the industrial loader manufacturers, Bray Construction Equipment Limited of Feltham, Middlesex.

After the introduction of the uprated Fordson Power Major in 1958, and the more refined Super Major with depth control and differential lock in 1960, Roadless conversions were sold in even greater numbers. In the last three months of 1959 alone, nearly 100 tractors were fitted with the four-wheel drive conversion for the Ford Motor Company in Britain. This was followed by a batch of just over 130 machines dispatched from Hounslow during January and February 1960, to the Ford Motor Company in Helsinki for use in Finland – one of Roadless's best export markets.

To increase the capacity of the Roadless four-wheel drive Super Major for industrial applications, the company introduced a new planetary hub reduction axle in 1963. This heavy-duty unit was based on the GMC 'wide axle' fitted with

ABOVE:
Roadless Super Dexta owned by Roger Haynes.

LEFT:
A view of Roadless's yard in Hounslow taken from Gunnersbury House in 1964. The photograph shows several 6/4 Ploughmasters awaiting dispatch, and a number of Super Majors for conversion. A Roadless Land-Rover can just be seen in the foreground.

ABOVE:
The inside of the Roadless assembly shop in 1964.

BELOW:
Built in 1962, the prototype Roadless Ploughmaster 6/4 is seen on trial with a five-furrow semi-mounted plough.

planetary hubs. Each hub featured a sun pinion gear running in constant mesh with five planet gears. The new axle was demonstrated on a tractor equipped with a Whitlock Dozaloda front-shovel. It appears that only two more 4WD Super Majors were fitted with the planetary axle, both going to C. H. Christiansen of Denmark to be used as skid-units for industrial loaders. A few Roadless Super Majors were also fitted with Brockhouse torque converters for industrial use between 1961 and 1962.

The Ford motor Company introduced the little 32 bhp Fordson Dexta in 1958, aimed at the smaller farmer. Before long, Roadless were receiving requests for a four-wheel drive conversion for this tractor from hill farmers and market gardeners who wanted the advantages of a light and compact machine with increased performance and traction.

The Roadless Dexta was launched at the 1959 Smithfield Show. Roadless did not manufacture the conversion for this tractor themselves, as they felt it was not a viable proposition to tool-up for what was expected to be a very small-scale production run, and bought the four-wheel drive parts in from Selene. The four-wheel drive system was based on a similar transfer box arrangement as used on the Major, but with a split-casing axle from a Dodge truck. The conversion became available for the Super Dexta from 1962, but in the end only about seventy Roadless Dextas and Super Dextas were sold, one of the last going to company chairman, G. E. Liardet, in 1964. As a price comparison – in 1962 the Roadless Dexta cost £1,101, which was not cheap; the Roadless Super Major seemed a lot more tractor for the money at £1,230.

Roadless could also see that something more powerful than the 52 hp four-cylinder Super Major was needed for the top end of its range of four-wheel drive tractors. The company had already used the Ford 590E six-cylinder industrial diesel engine in 1961 for a batch of eleven two-wheel drive Super Major tractors built for working in high altitude regions, and it was a natural progression to fit this engine to a four-wheel drive tractor. An experimental prototype, registration number 7271 MD, was completed in March 1962. This machine used the 590E engine connected by an adapter plate to a Super Major transmission. The transmission was modified to take the increased power by fitting stronger bearings in the gearbox, heavy-duty rear axles and a 13 in. clutch. Four-wheel drive was by the normal transfer box arrangement. The tractor went into production the following year with an improved front-axle mounting, and was sold as the Roadless Ploughmaster 6/4 rated at 76 bhp.

In designing the Ploughmaster 6/4, Roadless had taken into account power to weight ratios, and moved the front axle back to give the tractor a short 75 in. wheelbase with equal weight distribution over both axles, thus eliminating the need for additional weights. The first few built were painted in a green and yellow colour scheme. This attempt by Roadless to emphasise the new model and make it stand out backfired somewhat as the green paint had been applied over the original Fordson blue which started to wear through making the tractors soon look shabby. The bulk of the tractors were finished in the normal blue livery, although some were later given grey bonnets. Exactly two hundred 6/4s were built before the model was superseded by the Ploughmaster 90 in 1965, including one which was sent to Southern Rhodesia to operate as an aircraft tug with Central African Airways.

ABOVE:
The production model Ploughmaster 6/4 in 1963. The first few of these tractors were finished in a green and yellow colour scheme.

LEFT:
This Roadless Ploughmaster 6/4, purchased by a Northamptonshire farmer in 1964, had completed 4,000 hours work by 1967.

An interesting experimental version of the 6/4 was made in 1964, fitted with a Lucas T100 hydrostatic transmission. The tractor was developed in conjunction with Lucas Industrial Equipment, and had to have the engine uprated to 90 hp to cope with the power loss through the

Roadless

RIGHT:
An industrial Roadless Ploughmaster 6/4 delivered to Central African Airways operating out of Salisbury, Southern Rhodesia. The tractor was heavily ballasted to enable it to tow aircraft, such as the VC10 in the background which weighed over 100 tons.

BELOW:
The prototype Ploughmaster 6/4 at the end of its working days. Note the different front axle mounting to that used on the production models.

RIGHT:
Roadless Ploughmaster 6/4 at work with a four-furrow Lemken reversible plough.

hydrostatic transmission. The machine was unveiled to the public as the Roadless Super 90. Only one or two were made for experimental use. Plans to build further examples were suspended by the introduction of the Ford 5000, as the costs involved in developing a unit to fit the new design of tractor were considered too prohibitive. The option was offered for the new Ploughmaster 65 at the 1964 Smithfield Show, but none were ever made. Sadly, probably the only remaining example of the Super 90 is known to have been broken up in the last ten years.

Another batch of twenty-two six-cylinder two-wheel drive tractors was

transfer box between the gearbox and rear transmission. To overcome this, a side-drive unit was fitted over an aperture in the side of the main gearbox housing. The front axle was almost identical to that used on the Fordson and based on a GMC unit.

After consultations with Roadless, Selene, who had no designs on the B-450, agreed that International could market the four-wheel drive tractor anywhere in the world, including Italy, without restrictions. The built for the Ford Motor Company in Mexico at the end of 1964. These tractors were sold as the Roadless 6/2, and were again required for working at higher altitudes where the increased horse-power of the six-cylinder engine compensated for loss of power due to lower oxygen levels. A diesel engine operating at a height of about 10,000 ft was found to only deliver two-thirds of its maximum power.

Not all Roadless's four-wheel drive tractors were based on Ford skid-units. In 1961, the company was approached by International Harvester of Great Britain who required a four-wheel drive version of the 55 hp International B-450 tractor. The first prototype B-450 with Roadless four-wheel drive was completed at Hounslow in May 1961, and sent to International's factory at Doncaster for evaluation. Unlike the Fordson, the design of the International made it impossible to insert a

ABOVE:
Roadless Ploughmaster 6/4 working with a twin-leg subsoiler in 1965.

LEFT:
The Roadless Ploughmaster 6/4.

BELOW:
The Roadless Ploughmaster 6/4 Super 90 fitted with the Lucas hydrostatic transmission. Roadless's field sales operations manager, Ron Young, is at the wheel.

ABOVE:
The first International B-450 to be fitted with the Roadless four-wheel drive conversion is pictured at the yard in Hounslow in 1961.

RIGHT:
The production model International B-450 4-Wheel Drive built with kits supplied by Roadless from 1963 to 1970.

RIGHT:
International B-450 with Roadless four-wheel drive.

sales literature.

Drawings were made for a four-wheel drive conversion for a Nuffield Universal in late 1957. These were followed by plans for a Nuffield Universal four-wheel drive industrial tractor the following year. One four-wheel drive axle assembly was dispatched to Morris Motors in June 1959, but no details exist of any Nuffield tractors using Roadless 4WD equipment. At this time, the Nuffield Organisation was experimenting with its own four-wheel drive tractor using several different drive axles, and it is believed that the Roadless axle was used as part of this experiment. Roadless's stock book shows that the axle was returned to Hounslow in December 1961.

Another four-wheel drive project undertaken at Hounslow during the late 1950s and early 1960s, was the development of the Roadless Land Rover. In 1959, the Forestry Commission was experiencing difficulties with conventional Land Rovers getting stuck on the rutted forest tracks or grounding on fallen trees. The machinery research officer for the Commission, Colonel Shaw, suggested fitting the same 10 in. x 28 in. wheels to the Land Rover as Roadless had used on B-450 with Roadless four-wheel drive equipment was put into production, and the first twenty-five were built at Hounslow in October 1963. All the subsequent tractors were manufactured at Doncaster with Roadless supplying the four-wheel drive conversion kits. The tractor was sold by International as the B-450 4-Wheel Drive and remained in production until 1970. No mention was made of Roadless's involvement by International Harvester's

LEFT:
Roadless 109 Land Rover sold to the Central Electricity Generating Board for use in Wales in 1964.

its high-clearance front-axle conversion for the E1A Major. This worked up to a point, but the turning circle was dreadful and the gear ratios were now too high.

The Forestry Commission decided to develop the Land Rover further and sent a vehicle to Hounslow for Roadless to effect a proper conversion. Roadless retained the original transfer box and gearbox, but fitted Studebaker axles with GKN-Kirkstall planetary hub reductions and

LEFT:
Another view of the CEGB's Roadless 109 Land Rover working in Wales.

RIGHT:
The first prototype Roadless Land Rover which was supplied to the Forestry Commission for trials at Alice Holt Forest.

BELOW:
The second prototype Land Rover which was tested by the Rover Special Projects Department.

10 in. x 28 in. tyre equipment. This combination gave the vehicle a top speed of 30 mph. The front-axle was 14 in. wider than the rear axle to allow a decent steering lock. Roadless also uprated the brakes and suspension. The prototype was sent to the Forestry Commission's experimental station at Alice Holt for trials.

Sensing that the vehicle might be worth developing into a commercial product, Roadless built a second prototype and sent it to the Rover Special Projects Department who thoroughly tested it on the Motor Industry Research Association cross-country course at Lindley. Both vehicles performed well, but Roadless had to effect some modifications to the suspension, alter axle clearances and strengthen the chassis frame. After two years of testing, the Rover Company gave Roadless approval to sell the vehicle, which was offered from 1961 as the Roadless 109, with petrol or diesel engine options.

The petrol engined Roadless 109 cost £1,558 in December 1961, and the diesel version was listed at £1,658. A pick-up body for the vehicle was £172 extra. It was an expensive machine and very few were sold. One went to the Central Electricity Generating Board, and this vehicle is now preserved by the Dunsfold Land Rover Trust; another was sold to Scotland for retrieving deer. One was sold to a sheep farmer in the Falklands, and is reputed to be the first vehicle commandeered by the

LEFT:
One-wheel drive – pedal power. A prototype Roadless bicycle trailer, designed primarily for transporting churns of rubber latex in Malaya, is tested to the limit by two of Roadless's burly fitters.

Argentineans during the Falklands conflict, and another was exported to Gibralter. The Institute of Hydrology bought two to be used in Wales, one of which is still at work. There is also evidence of one ending up in Australia with the Snowy Mountain Authority. It is believed that no more than nine were built between 1961 and 1964. Roadless's own demonstrator was the only vehicle to be fitted with a diesel engine. This was not sold until 1969, when it went to a hill farm in Wales and is thought to be still in use.

Even with everything else that was going on at this time, Roadless still found time to play about with a few other projects. However, a bicycle trailer brought out in 1960 was sadly one of the last small-scale developments that the company allowed itself as four-wheel drive production increased.

Four-wheel drive tractors had now become the mainstay of Roadless Traction's business, and the Fordson conversions had been a particular success for the company. By the time manufacture of the Super Major came to an end at Dagenham in 1964, around 3,000 Roadless 4WD Fordsons were at work across the world. Ford Tractor Operations had a new range rolling off the production lines at the recently built Basildon factory; with four-wheel drive conversions for these new models in the pipeline, Roadless could look to the future with optimism.

CHAPTER 5
Power on the Land
1965 to 1975

A comparison in height between the standard Roadless 115, and the high clearance version which was designed for sugar cane work in Puerto Rico.

*BELOW:
Rupert F. C. Booth, managing director of Roadless Traction from 1962 to 1977.*

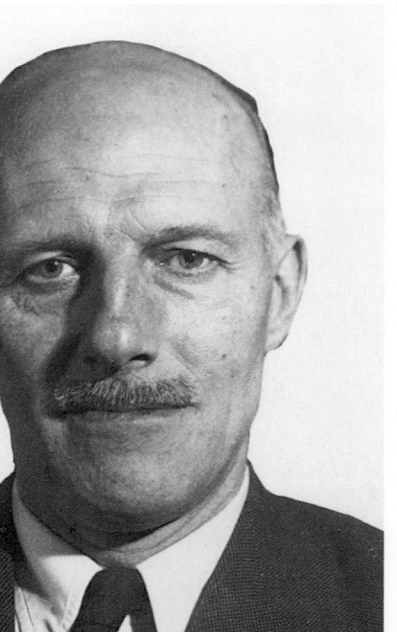

*RIGHT:
Ven Dodge, sales manager at Roadless from 1964, lectures a group of students from Merrist Wood Farm College on the finer points of the company's tractors.*

1965 saw the beginning of a new era for Roadless – and the end of an old one. A new range of Ford tractors was flowing out of Basildon, and Roadless was working on fresh designs of four-wheel drive models, based on these new tractors, that would see the company's sales boom over the next ten years. The four-wheel drive tractor was here to stay. Roadless had played a big part in its establishment in Britain – and was ready to secure its future development into the '70s.

Crawler and half-track sales had been in decline, and manufacture had ceased as four-wheel drive production took over at Hounslow. The days of Roadless tracks were long gone. This era finally closed with the passing away of Lieutenant-Colonel Philip Johnson, who died on 8 November 1965, aged eighty-eight.

Even in his old age, Philip Johnson had remained active within the company, and was still a director at the time of his death. He had never really retired and was undertaking business trips abroad right up to nearly the last years of his life. He had remained managing director and chairman of Roadless until 1962, relinquishing the former position to Rupert F. C. Booth, a qualified civil engineer, and son of Cecil Booth who had joined the board of Roadless in the 1920s. The chairmanship was taken over by Mr. G. E. Liardet, formerly chairman of the Simms group. Charles Skelton, now in his seventies, remained works director until his death in 1973. The job of works manager then went to Wally Hawkins who had been working under Skelton. The other director and largest shareholder at this time was Miss Doris Ethel Jones, daughter of the first chairman, Sir William Jones. The company secretary was Mr. A. V. Dunworth, who had joined Roadless in the 1950s.

A new sales manager had been appointed to the company in 1964, in the form of Alfred Ventham Dodge. 'Ven' Dodge, previously sales manager of the Ford dealers, Crimble of Staines, had already sold and demonstrated many Roadless Fordsons across the Surrey area and was very familiar with the company's products. He had enjoyed particular success with the Roadless Dexta, which he had found to be the ideal product to offer the county's many market gardeners against the Massey Ferguson 35.

In 1966, Leonard Tripp retired at the age of 65, and Vic Crockford was appointed as the new chief designer. Crockford was a very accomplished designer who could offer Roadless wide experience in a variety of fields. After a spell in the aircraft industry, he had worked both in Holland and for a London consultancy firm. He had been involved with hydrostatic aircraft handlers, and the design of both the mobile launcher for the Bristol Bloodhound rocket, and

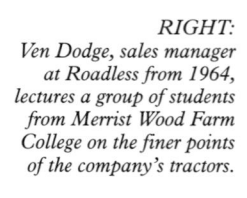

LEFT:
The first Roadless Ploughmaster 65 which was built 'down the line' at Basildon in a rushed attempt to get it ready for the 1964 Smithfield Show.

BELOW:
Roadless Ploughmaster 65 photographed at Hounslow in 1965.

the mechanical side of the flight recorder for the ill-fated TSR-2 project.

Vic Crockford brought fresh ideas to the company, and some of his revolutionary designs were to put Roadless towards the forefront of four-wheel drive development. With Crockford designing the machines and Dodge selling them, people were often heard to remark on the company's unique double act of 'Vic and Ven – the Roadless Men'. The line-up at Roadless was further strengthened by the addition of Arthur Battelle, who joined as service manager from Ford in 1967. The new team was now in position to take the company forward into the next decade.

Most Roadless four-wheel drive tractors built from this time were based on Ford skid-units. To understand the

ABOVE:
Roadless Ploughmaster 65 at work with a set of heavy discs.

RIGHT:
Roadless Ploughmaster 65 pulling a set of Lundell seed drills.

development of the Roadless, it is necessary to look at the evolution of the Ford.

Known as the 6X range, the first of the new Ford tractors were built at Basildon in the autumn of 1964, and were unveiled to the public at the Smithfield Show in December. Unlike previous Ford models, the 6X range had evolved as a standardised range of virtually identical tractors that would be produced from all the Ford factories in different countries for world-wide distribution – the advantage to the conversion builders, such as Roadless, being that their overseas sales would benefit from an already established global parts and service network.

Four new models were released from Basildon: the three-cylinder 2000, 3000 and 4000 tractors, and

LEFT:
Roadless Ploughmaster 65 used for direct nitrogen injection.

the top of the range four-cylinder 5000. The three largest tractors had the option of eight-speed manual gearboxes or the ten-speed Select-O-Speed epicyclic transmission. Of immediate interest to Roadless was the 65 hp 5000, as a four-wheel drive version of this tractor was an ideal successor to the top-selling Roadless 4WD Super Major.

The new Roadless model became known as the Ploughmaster 65. Designed to the same conventional layout as the previous Roadless four-wheel drive tractors, it was fitted with a transfer box driving a GMC front axle. In the company's rush to get the new tractor ready for the 1964 Smithfield Show, the four-wheel drive conversion was sent to Ford and fitted 'down the line'

BELOW:
A batch of Ploughmaster 65 tractors lined up at Hounslow ready for export to the West Indies.

RIGHT:
The first Roadless Ploughmaster 90 photographed at Hounslow in June 1965.

RIGHT:
A 1966 Roadless Ploughmaster 90 cultivating in Lincolnshire.

at Basildon. The tractor then appeared on the Roadless stand still sporting Ford badges as there had been no time to get the Roadless name on to the bonnet. The new machine was priced at £1,587.

Roadless also announced details of the imminent release of a new six-cylinder model to replace the 6/4 Ploughmaster. Designated the Ploughmaster 90, the first of these new six-cylinder tractors was not actually built until June 1965. It was also based on the Ford 5000 skid-unit, but continued to use the old 590E

LEFT:
Roadless Ploughmaster 90 on demonstration as an industrial prime mover at Felixstowe docks.

power-plant, now rated at 90 hp. Both the new Roadless tractors had shorter wheelbases than the equivalent Ford two-wheel drive models. The front axles were mounted further back to give excellent weight distribution and to ensure that all the power was converted into traction.

The tractors were capable of hard work, but some were worked much harder than had been expected, giving Roadless the odd headache. A few Ploughmaster 90s developed front-axle and hydraulic problems, and the company had to replace several front differentials on this model. One 90, fitted with a British Twin Disc torque converter and sold to DFDS at Felixstowe for

LEFT:
A Roadless Ploughmaster 90, fitted with a British Twin Disc torque converter, used by DFDS for moving containers at Felixstowe docks.

RIGHT:
Roadless Ploughmaster 95 introduced in June 1966.

RIGHT:
A 1966 Roadless Ploughmaster 95 ploughing in Sussex.

Power on the Land

LEFT:
Roadless Ploughmaster 95 at work in 1967. The cab was built by Fritzmeier.

BELOW:
The turbocharged Roadless 80 pan-busting stubble.

BOTTOM:
A Roadless Ploughmaster 95 fitted with a GMC-based planetary axle and oversize tyres.

moving containers on and off the ferries, broke in half twice. After the second time, a somewhat alarmed Vic Crockford went to investigate only to find that the tractor was being used to move large semi-trailers loaded with massive boiler shells. The semi-trailers carrying the boiler shells were lower than the normal semi-trailers used at Felixstowe, and the driver was coupling the fifth-wheel dolly to the front of the tractor and, after tipping the engaging plate at an angle, using it to ram the dolly under the trailer. This caused a high compressive shock load to pass through the tractor, fracturing the adaptor casting between the engine and torque converter transmission. The 90 was only built for a year before being replaced in June 1966 by the Ploughmaster 95, fitted with the more powerful new 2703E engine, a heavy-duty front differential, modified hydraulic linkage and an assister ram.

From early 1966 onwards, Roadless offered the Ploughmaster 46 for the smaller farmer. Based on the three-cylinder 46 hp Ford 3000, this tractor was fitted with a version of the four-wheel drive system used on the Dexta, but using different ratios. As with the Dexta, the conversion was not built by Roadless, but bought in from Selene. In July 1967, the Ploughmaster 46 cost £1,509. The conversion kit could be bought separately from £560. The Ploughmaster 65 was now £1,805, and the six-cylinder Ploughmaster 95 cost £2,525.

Experiments in 1966 with a Ploughmaster 65 fitted with a turbocharger led to the introduction of a fourth model to the Roadless range – the Ploughmaster 80. Built for a short while between 1967 and 1968, it was based on a

RIGHT:
The Roadless Ploughmaster 46, based on the Ford 3000.

BELOW:
An exploded view of the turbocharged Roadless Ploughmaster 80. (Power Farming copyright).

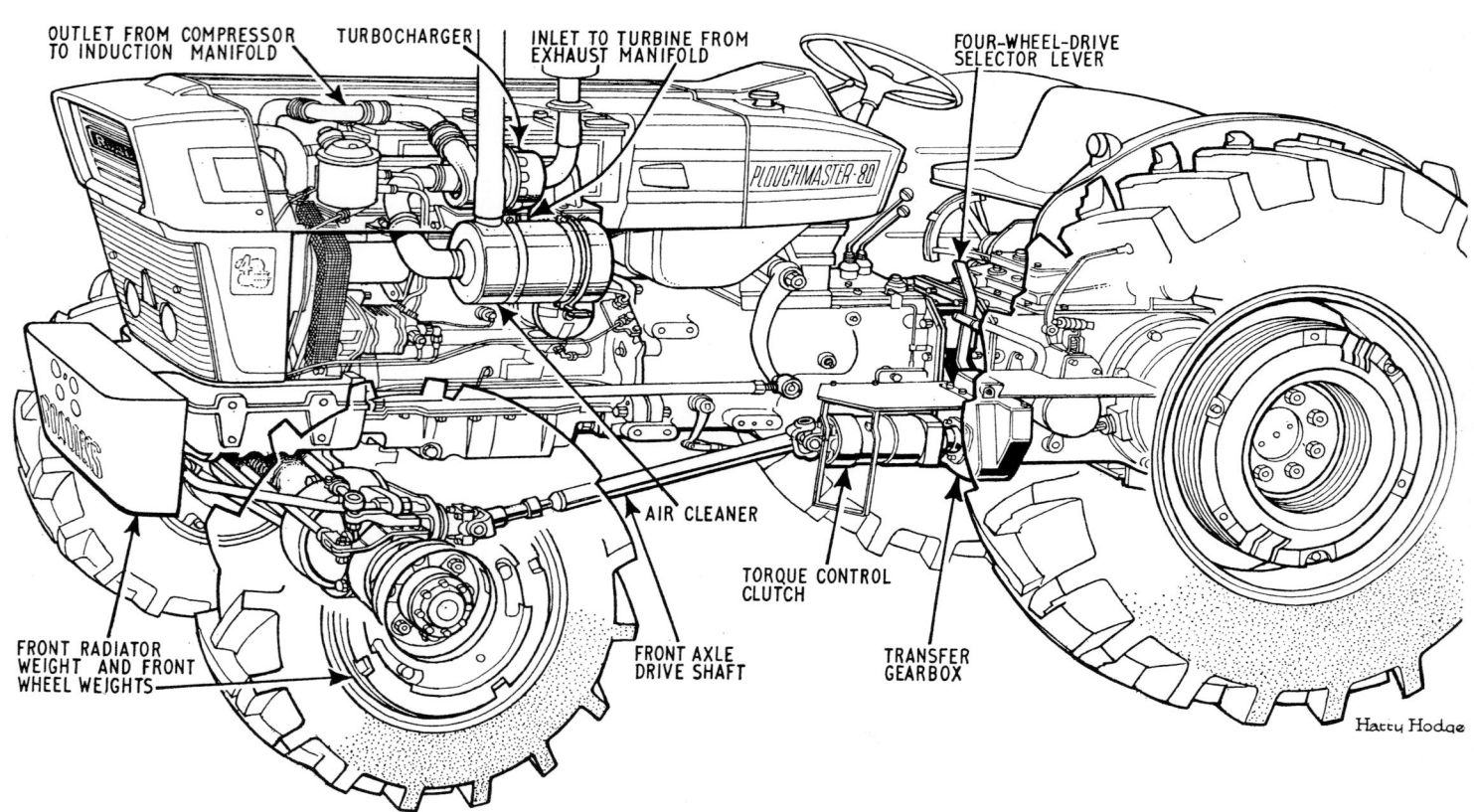

Ford 5000 fitted with a CAV turbocharger which boosted the available power to 80 hp. This tractor had the same heavy-duty differential in the front axle as the 95, and also had an assister ram. It was priced at £2,165.

The engine fitted to the 80 was not strengthened for turbocharging, but Roadless avoided any overheating or oil consumption problems by fitting a larger capacity radiator and an oil-cooler. Turbocharging was still in its infancy when this tractor was made, and many farmers shied away from the concept, preferring the six-cylinder engine for power. In the end, just over twenty were built. County achieved no greater success with its 854T model which used the same engine and turbocharger.

Of the four Roadless tractors now on offer, the Ploughmaster 65 was by far the most popular. It enjoyed considerable export success and many went to Trinidad, including one batch of twenty-eight going out to the local dealers, Caroni Limited, in late 1966. Trinidad became a good export market for Roadless. The sugar cane grown there was cut by hand over a six-month period, and the harvesting had to carry on through the wet periods. Because of the nature of the terrain, four-wheel drive tractors were the only machines capable of hauling the loads of cane. Eventually Roadless had over 100 tractors working in the country. The sugar cane was grown on hillsides in what was known as

TOP:
An industrial Roadless Ploughmaster 65 fitted with a torque converter transmission and supplied to Carlsberg for moving crates of lager from the docks to the warehouse in Goole.

ABOVE:
An industrial Ploughmaster 65 with towing equipment built for Laker Airways.

LEFT:
Laker Airways' Ploughmaster 65 towing tractor in use at Gatwick airport.

Roadless

RIGHT:
The Roadless 700 was fitted with a GMC-based planetary axle and a Steel-Fab loader.

penny-plain formation, and the tractors had to stand the rigours of bringing loads of cane down the steep slopes, and negotiating a deep drainage ditch before reaching the road. To cope with the work, Roadless fitted many of the tractors with specially reinforced front axles. Even as production of the Ploughmaster 65 came to an end, the last twenty built were supplied to Tate and Lyle for use in Trinidad in August 1968.

A few Ploughmaster 65s were built for industrial use, including

RIGHT:
The 1968 Roadless 700 loader, based on a Ploughmaster 65 fitted with a Brockhouse torque converter.

RIGHT:
Close-up of the transmission on a later Roadless 700 based on a Ploughmaster 75. The forward and reverse shuttle lever for the Brockhouse torque converter transmission can be seen on the right of the main gear levers.

an aircraft tug which was sold to Laker Airways and operated at Gatwick. One or two 65s were also fitted with British Twin-Disc torque converter transmissions and used for heavy haulage, including one sold to Carlsberg. This tractor was used in Goole, North Lincolnshire, to haul crates of lager from the docks to the warehouse five miles away. Carlsberg was allowed special dispensation for the 65 to pull its heavy load through the town centre provided it was preceded by a man on a bicycle waving a flag.

Built in 1968, the Roadless 700

industrial loading shovel was another interesting development. Based on a Ploughmaster 65, it was fitted with a GMC-based planetary hub heavy-duty front axle, and a loader supplied by Steel-Fab of Cardiff. The normal flywheel clutch was replaced by a Brockhouse torque converter which gave forward and reverse drive in all gears. The standard eight-speed Ford gearbox was modified to give six speeds after removal of the two lowest ratios and reverse gear which were no longer necessary.

ABOVE:
A Roadless Ploughmaster 65 used by the Caroni Sugar Company in Trinidad hauls a trailer loaded with two 1-ton bundles of sugar cane.

LEFT:
Cane harvesting in the West Indies with a Roadless Ploughmaster 75.

BELOW:
Roadless Ploughmaster 95 working with a heavy-duty cultivator.

export markets, especially Trinidad, and a batch of thirty two tractors went out to Tate and Lyle in 1969. Several were also equipped for industrial use, and some were fitted with torque converter transmissions.

1968 was the year Roadless built its first equal-sized four-wheel drive tractor. For several years the company had worked on developing a heavy-duty equal-wheel machine to compete with the products of its nearest rival, County. The problem in designing this type of tractor was to

The 700 was also available without the loader as a prime mover. Originally only one was built, but another six, based on the later Ploughmaster 75, were supplied to Israel between 1969 and 1972.

The Ploughmaster 75 replaced the 65 when the improved Ford Force 6Y range came out in 1968, and the 5000 was uprated to 75 hp. The six-cylinder Ploughmaster 95 and the 3000-based 46 also gained the benefit of the new Ford styling, but the short production run of the 80 had ended by this time and none were built on Ford 6Y skid-units.

The 75 proved to be as popular as the old 65. Again it found favour with the

RIGHT:
Roadless tractors based on the Ford Force 6Y skid-units lined up outside Gunnersbury House. Two Ploughmaster 46s are at the front. The one on the right is fitted with oversize tyres for use with the Forestry Commission.

BELOW:
Roadless Ploughmaster 95 fitted with a Duncan safety cab.

provide an arrangement for mounting the front differential that did not require the engine elevating above the front axle and so losing the advantages of a low centre of gravity. County had got round the problem by not using a front differential and having two propellor-shafts, one to drive each front hub. Northrop, on the other hand, had resorted to raising the height of the tractor to accommodate the front axle under the engine, and was hampered by a high centre of gravity.

Roadless's first designs for an equal-wheel machine were drawn up in the early 1960s by Leonard Tripp. Tripp's idea was to place a spacer between the engine and gearbox with an extended input shaft. The spacer was cut away to allow room for an in-line front axle and differential to be mounted underneath it without needing to elevate the engine. The design was similar to that used on the Hungarian Dutra four-wheel drive tractor, but was by no means perfect. The main disadvantage being that with the whole engine mounted forward of the front axle, the machine would be very long with too much front overhang

122

which would make it ungainly and awkward for turning and working in tight spaces.

After Vic Crockford was appointed chief designer for the company, he took a novel approach to the equal-wheel concept and drew up plans for a completely new front axle that was both revolutionary and simple in its design. A conventional differential was used with the half-shafts extending to two auxiliary reduction gearboxes which lifted the drive-line up and allowed large diameter wheels to be fitted without the need to elevate the engine and lose the low centre of gravity. With this arrangement, the differential centre line was nearly 7 in. below the front wheel centres. The axle also had the advantage of three separate gear-reduction stages. The first stage was the differential, and the second stage was the reduction gearboxes. The third stage was provided by fitting planetary reduction front-wheel hubs.

ABOVE:
A Roadless Ploughmaster 75 discing land for sugar cane in St Kitts.

LEFT:
The prototype Roadless 115 at work with Sussex farmer Richard Place.

Drive to the front axle, which also had the benefit of a differential lock, came from the conventional Roadless transfer box system.

Another feature of the new axle was its weight. Fabricated with heavy cast steel axle supports, the combined front transmission weighed in at a hefty 1 ton. Having most of this weight below the front wheel centres contributed towards a very low centre of gravity, and excellent stability. More importantly, with 6,600 lb of the tractor's 11,000 lb gross weight over the front axle when static, Roadless had achieved near perfect weight distribution without the addition of extra weights – a feat that few other manufacturers could match.

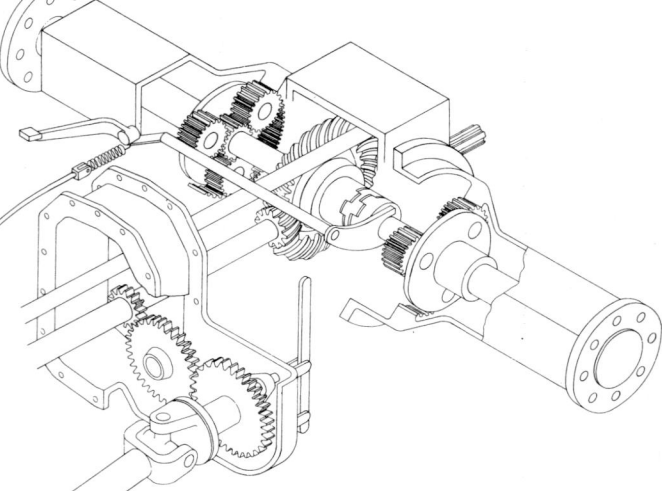

RIGHT:
The four-wheel drive transmission arrangement for the Roadless 115.

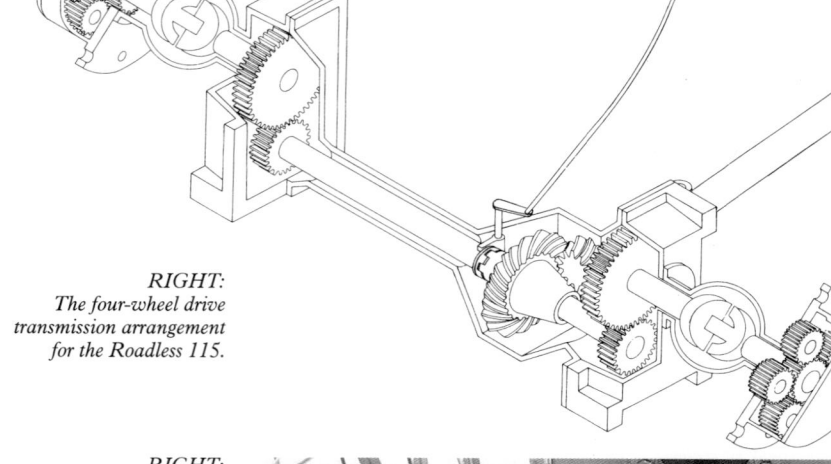

RIGHT:
A worm's eye view of the Roadless 115 showing the front axle mounting arrangement.

Roadless's first equal-size four-wheel drive tractor was ready in prototype form by February 1968. The axle on this machine used a Kramer differential and Bray planetary front hubs. The Ford-based tractor was powered by a six-cylinder 2704E engine developing 115 hp (gross). The gearbox was strengthened with high capacity bearings, and an extra-heavy-duty clutch was fitted. Hydrostatic power steering completed the specification list.

The prototype was loaned to a Sussex contractor, Richard Place of Framfield, for field trials. Mr Place ran a couple of Dutra tractors, and so the Roadless was painted the same colour red and fitted with a Fritzmeier cab to disguise it from a distance. Consequently, Roadless employees often referred to the tractor as 'The Red Devil'.

During its first 1,000 hours work, the prototype only lost twelve hours due to breakdowns, which all proved to be minor faults, and before it was fifteen months old, it had an impressive 2,300 hours on the clock. Its workload had included covering over 3,000 acres with a set of 17 ft discs at a rate of up to nine acres an hour; ploughing 900 acres with a five-furrow 14 in. Bamford Kverneland plough, at one time achieving three acres per hour, and cutting nearly 3,000 tons of silage with a Gehl forage harvester, handling 130 to 160 tons a day.

Designated the Roadless 115, the production model equal-size four-wheel drive tractor made its debut at the Royal Show in July 1968, priced at £3,525. Richard Place had been so impressed with the prototype, that he bought both it, and a production model 115, to replace his two Dutras. The prototype, after an overhaul at Hounslow, was returned to him in August 1969. The production model 115s had revised styling to match the new Ford Force range. An extended radiator shroud was available to accommodate an extra front-mounted fifteen gallon fuel tank, and an assister ram was fitted giving the tractor a lift capacity of 6,000 lb at the lower link ends. The differential and planetary hubs were now of Roadless's own design.

A high clearance version of the 115 was introduced in January 1969 for sugar cane work. The conversion was achieved by inverting the front axle so that the reduction gearboxes were turned down. The rear end was raised by using drop-boxes consisting of heavy-duty casings containing a vertical train of gears. These rear drop-boxes were bought in from County who built them for their High-Drive conversions. Over forty high clearance 115s were supplied up to 1976, all going to Clemente Santisteban for use in Puerto Rico. Two high clearance 75s were also made in 1971.

Another interesting version of the 115, with an extended chassis for hillside ploughing was made for the Forestry Commission. The tractor was fitted with a reduction box, and the front axle was moved as far forward as possible by modifications to the A-frame to allow the front wheels to climb out of gulleys.

A few 115 tractors were also sold to the Bruff Manufacturing Company of Suckley, near Worcester, for fitting with McConnel-Thornton-Garnett trenchless drainers for laying plastic or

ABOVE:
The prototype front axle for the Roadless 115 featuring Bray hubs and a Kramer differential. Note the Ford skid-units ready for conversion lined up in the background.

LEFT:
The heart of all Roadless four-wheel drive tractors – the transfer box ready to be mounted between the gearbox and rear transmission.

BELOW:
The prototype Roadless 115 'Red Devil' ploughing in Suffolk.

ABOVE:
A production model Roadless 115 ploughing 'on the land' with a seven-furrow SK plough.

RIGHT:
A 115 under test at Hounslow on an air brake dynamometer.

Power on the Land

LEFT:
Chief designer, Vic Crockford, at the wheel of a production model Roadless 115.

BELOW:
The extended radiator shroud on this Roadless 115 housed an auxiliary fifteen-gallon fuel tank.

BOTTOM:
Roadless 115 at work with a Bamford Kverneland plough.

site. With sales booming, Roadless had no plans to sell the business, and was considering its options on the site. At one time, an approach was also made by Matbro with a view to buying the company, but for different reasons. The Surrey-based manufacturers of fork-lifts and materials handling equipment were looking to extend their product base and possibly expand further into the agricultural market. Negotiations were opened, but no agreement was reached.

In 1970, both Roadless and its tile drains. Marketed as the Bruff TG3, the machine was designed for one-man operation. When draining, the unit was powered by a hydrostatic winch pulling against a land anchor, and the four-wheel drive was only used for moving into position, and from site to site. The front-mounted winch was driven by two low-speed radial hydrostatic motors.

Not all the company's products were Ford-based. International Harvester were still customers for Roadless axles, and over 180 conversions were supplied to Doncaster in 1970 alone. The B-450 was still in production, but Roadless had now supplied conversions for two other models of International tractor. A prototype four-wheel drive version of the 62 hp B-614 was built at Hounslow in August 1966, and fifty conversion kits went to the International factory in 1968. Roadless supplied considerably more conversions for the 66 hp 634, which replaced the B-614 in late 1968. The Roadless version, of which over 200 were made, was marketed as the International 634 Four-Wheel Drive, as opposed to the International 634 All-Wheel Drive, which was an equal-wheel machine, built using parts supplied by County.

By the end of the 1960s, Roadless had become established as a leading player in the four-wheel drive field. Sales were flourishing as the demand from farmers for increased power and traction continued to rise. The export market was also buoyant, and Roadless tractors were sent out to Africa, the West Indies, South America, the USA and many other countries.

Roadless had grown, but so had Hounslow. The land was now as valuable as the company, and Roadless attracted takeover bids from property development firms keen to acquire the

RIGHT:
Demonstrator, Bruce Keech, at the wheel of a Roadless 115.

RIGHT:
Roadless 115 and Ploughmaster 75 tractors, built in 1969 for export to the Vaux Tractor Company in the USA.

rivals, County, were working on similar new designs of four-wheel drive tractors of unequal-size wheel configuration, and correspondence between Rupert Booth and Raymond Tapp of County shows a planned merger of interests. Falling short of integration of the two companies, the anticipated line of co-operation included an exchange of information on designs, and the proposed marketing of a uniform range of unequal-wheel tractors for export under the County-Roadless label. County would provide the 4000-Four model, and four-wheel drive versions of the 5000, and the new 7000 tractor which was under development at Ford, would come from Roadless. The ideas seemed sound, but were never sanctioned at board level, and County and Roadless remained independent.

The same show saw the introduction of another new Roadless tractor – the 94T.

The four-cylinder 94T, based on the recently launched 94 bhp turbocharged Ford 7000 tractor, featured a new diff-locked planetary hub reduction front axle of Roadless's own design and manufacture – an important step in the evolution of the company's own axles. This new heavy-duty axle was designed by Vic Crockford using some of the design features incorporated in the 115 axle. The first version of the axle used Selene planet gears in the reduction hubs, and had a fabricated steel casing to house the differential. Because of the distinctive shape of this casing, the unit was often referred to as the 'coffin' axle. This Mark 1 axle was only intended to be a

ABOVE:
One of two high clearance Roadless 75 tractors built in 1971 for Clemente Santisteban for sugar cane work in Puerto Rico.

LEFT:
This view of the high clearance 75 illustrates the tractor's 21 in. ground clearance. A cane ridger can just be seen mounted on the rear of the machine.

BELOW:
One of over forty high clearance Roadless 115 tractors that went to Puerto Rico is used for ridging sugar cane.

The early '70s saw a few changes to the Roadless range. In September 1970, new regulations came into force which required all tractors sold on the UK market to be fitted with safety cabs. Roadless went to Alexander Duncan of Aberdeen for its cabs, which were to feature on all the company's tractors for several years.

In 1971, the Ford Motor Company upgraded its range of industrial engines, and the new units were immediately adopted by Roadless. The Ploughmaster 95 now had the uprated 2713E engine, while the 2714E unit powered the 115. The third new six-cylinder engine in the range, the 120 hp (gross) 2715E, was fitted to the 115 skid to produce a new model – the Roadless 120 which was launched at the Smithfield Show.

stop-gap unit; some trouble was encountered with the Selene planets breaking up, and it was quickly replaced by the improved Mark ll axle which had a cast steel casing and Roadless's own planet gears.

The double reduction gearing in the new axle had at least three times the torque capacity of the old GMC-based axle which was still used on all tractors designated 'Ploughmaster'. The new planetary axle was also available for the 95, which was then marketed as the Roadless 95, as opposed to the Ploughmaster 95 with the GMC axle. The 95 was replaced in 1974 by the Roadless 105 which used the 104 bhp 2714E engine. All the 105s had the new planetary axle, and the use of the GMC axle was now confined to the Ploughmaster 75. One of the Roadless planetary axles was also fitted to an American 100 hp Ford 8000 for Ford France S.A. in August 1973.

Production of both the smaller Ploughmaster 46, and the International 634 4WD had come to an end by 1972. However, that year Roadless brought out a four-wheel drive conversion for the 45 hp International 444 using an axle bought in from

TOP:
An extended version of the Roadless 115 for use with the Forestry Commission.

ABOVE:
Roadless 115 equipped for drainage work.

RIGHT:
Roadless 115 fitted with a McConnel-Thornton-Garnett TG3 trenchless drainer.

Power on the Land

LEFT:
The prototype International B-614 Roadless four-wheel drive conversion at Hounslow in August 1966.

tractors was never demonstrated more clearly than at the Long Sutton 'Tractors at Work' trials held in Lincolnshire during the 1970s. Tractors working at Long Sutton were measured in terms of output and efficiency – determined by relating the number of acres ploughed per hour to pto horse-power which was tested on a dynamometer. Roadless entered the Sige company in Northern Italy, and a few kits were supplied up to 1975. This conversion featured a multi-plate over-centre clutch in the transfer box which allowed the front-wheel drive to be engaged and disengaged on the move. A large number of the transfer boxes were sold to the French Manitou company who fitted them to the International 444 skid-units used on its rough terrain fork-lifts. Roadless did not supply drive axles as Manitou bought in the Sige units direct from the Italian company. Vic Crockford also designed a four-wheel drive kit for the larger 574 and 674 tractors, but only three were made. The last International to be fitted with Roadless four-wheel drive was a 45 hp 384 which went to Oslo in 1979. Drawings were also made for a four-wheel drive conversion for a David Brown tractor in 1975, but none were built.

The high performance of the Roadless the event seven times and came away with six first positions, and on two occasions the company secured second place as well. All the Roadless tractors at the Long Sutton trials were demonstrated by Mervyn Ford, who had joined the company at the age of twenty-one after serving an apprenticeship with Massey Ferguson. The invitation to join Roadless as demonstrator had come soon after meeting Ven Dodge at a farm near Cheltenham in 1969, where he had successfully demonstrated a two-wheel drive Massey Ferguson 1100 against a Roadless 115. Mervyn

ABOVE:
International 634 Four-Wheel Drive fitted with the Roadless conversion kit.

LEFT:
Roadless Ploughmaster 75 fitted with a Duncan safety cab.

RIGHT:
Roadless Ploughmaster 75 at work ploughing in stubble. This tractor has a slightly later Duncan cab.

RIGHT:
Roadless Ploughmaster 95 fitted with a Duncan safety cab.

Ford had the accolade of twice holding the world record for ploughing an acre plot in the quickest time. The first time was in 1968 with a Massey Ferguson 1100, but the second attempt was after the move to Roadless with a 115 in 1970. After a spell as a service representative under Arthur Battelle, Ford moved into sales and eventually became eastern area regional and export sales manager for the company. However, he still personally prepared and demonstrated the tractors for Long Sutton. Roadless took the Long Sutton demonstrations very seriously. Preparation of the tractor for the event took nearly a week. All the specifications were thoroughly checked, and the fuel pump was removed, tested and calibrated to the precise settings. Roadless usually fitted the tractors with Kleber radial tyres for the event, which would have the exact correct pressures. Even the plough came in for some

LEFT:
The Roadless 120 introduced in 1971.

LEFT:
Roadless 120 in action, ploughing near Ipswich, Suffolk, in March 1972.

LEFT:
Roadless 94T working on a one in three slope in Dorset.

Roadless

RIGHT:
Arthur Battelle in France with a Ford 7000 which he had fitted with a Roadless Mark l planetary axle.

attention, and the breasts would be highly polished.

The Roadless 115 came away with first position at both the 1970 and 1971 Long Sutton performance tests. The following two years, the company entered both a 95 and a 94T, each equipped with the new planetary axle. The Roadless 95 took top place, with the 94T coming in second, both years running. A remarkable testimony not only to the performance of the tractors, but also to the dedication of the personnel at Roadless. Long Sutton was a two day event, and after the performance tests held on the first day, Ven Dodge would work on the figures overnight to have a press release ready for the second day's attendance.

ABOVE:
Roadless 95 fitted with the Mark l planetary axle.

RIGHT:
Mervyn Ford demonstrates the Roadless 94T.

134

ABOVE:
Roadless 95 with the planetary axle working a three-furrow Ransomes TSR103 reversible plough.

LEFT:
Ford 8000 fitted with the Roadless planetary axle in August 1973. The conversion was done for Ford France S.A.

RIGHT:
International 444 fitted with a Roadless four-wheel drive conversion kit.

RIGHT:
International 444 with Roadless four-wheel drive photographed in the garden at Gunnersbury House in 1972.

BELOW:
International 384 with Roadless four-wheel drive supplied to Oslo in 1979.

Having won the first four Long Sutton tests, the company felt it had nothing further to prove in terms of performance or sales, and did not enter again until 1978. By the mid-1970s, Roadless's order books were full to the point of overflowing, and there was an eighteen to twenty-two month waiting list for the tractors. The company was unable to even replace its own demonstration machines, which by 1975 were in their fourth year of use, another factor which contributed towards the decision not to participate at Long Sutton.

Unfortunately, Roadless Traction had reached its zenith, and the last five years of the decade were to see change and decline for the company.

Power on the Land

*LEFT:
Roadless 94T at the 1972 Long Sutton ploughing demonstration.*

*LEFT:
Roadless 95 at the 1972 Long Sutton ploughing demonstration with a three-furrow Ransomes TS84 plough.*

*LEFT:
A Roadless 105 lined up with a 120 in Lincolnshire.*

The Roadless Logmaster forestry skidder at work.

CHAPTER 6
Change and Decline
1975 to 1983

*RIGHT:
Roadless 98
introduced in 1975.*

*RIGHT:
Roadless Ploughmaster 78
being used for extracting
timber.*

The launch of the revamped 7A range of Ford tractors in 1975, with improved engines and hydraulics, was to provide Roadless with uprated skid-units on which to base its models. However, Ford was expanding its range and intended to exploit every sector of the market, and in a few years was to produce four-wheel drive versions of its own tractors – a move which was to seriously affect Roadless and, in some part, contribute to the difficulties which lay ahead for the company.

The Ploughmaster 75, which had been one of Roadless's best selling models, was replaced by the Ploughmaster 78, based on the new 78 hp Ford 6600, in November 1975. The Roadless 98 superseded the old 94T the following month, and used the turbocharged 97 hp Ford 7600 skid. With the exception of a few more high clearance versions for Puerto Rico, no 115s had been made since May of that year. The more powerful 120 remained in production, but now incorporated the refinements introduced with the new Ford 7A range. The Roadless 105 was built for another six months, but a new replacement model was in the pipeline.

Roadless continued to use Duncan cabs, but new legislation required all agricultural tractors manufactured from June 1976 to be fitted with a quiet cab. Ford had introduced a new straddle cab made for them by GKN-Sankey, but this would not fit the Roadless tractors because of the transfer box. The Duncan Supercab was used for a time, but the company felt it was a little basic and turned to Lambourn Engineering of Newbury, Berkshire, for a new cab which became a standard fitment on most Roadless tractors. Export models, now mainly Ploughmaster 78s for sugar cane work in the West Indies, and some to Fyffes for use on banana plantations in Costa Rica, were still supplied without cabs.

Change and Decline

LEFT:
Roadless 98 equipped for use as a forestry tractor.

BELOW:
A line of Roadless Ploughmaster 78s destined for export to Barbados for sugar cane work.

The 105 was finally replaced by the new model Roadless 118 in June 1976. The 118 had the 114 bhp Ford 2715E engine, which together with the diff-locked planetary axle, made a powerful combination. The Lambourn Q-cab was fitted, and the tractor had the benefit of hydrostatic steering. It was first shown to the public at the Royal Show at Stoneleigh in July, priced at £11,030.

In 1977, Roadless introduced a flat-floor cab across its range, using the Lambourn LFQ-cab with modifications carried out at Hounslow. The tractors with the new cabs were announced in

LEFT:
Roadless Ploughmaster 78 hauling sugar cane in the West Indies.

June, coinciding with the week of celebrations to mark the Queen's Silver Jubilee, and consequently were named the J Series. Prices for the Ploughmaster 78 and the Roadless 98 were now £9,850 and £13,200 respectively. The J Series 118 cost £14,900, and the 120 could be bought for £16,200. Further improvements to the cab, including larger opening side windows and doors with gas stays, led to the introduction of the K Series range in late 1978, and another price increase.

Roadless returned to the Long Sutton 'Tractors at Work' demonstration in 1978. Unfortunately, Mervyn Ford drew a plot of heavy land, and his 118 only managed fourth place in the results. He was back on form at the following year's event held on land owned by J. Van Geest (Farms) limited at Saracens Head, near Holbeach. The 118 recorded 5.4 per cent, and once again put Roadless at the top of the tables in terms of acres ploughed per

ABOVE:
Taken in 1977, this photograph of a Roadless Ploughmaster 78 operating under wet conditions in Trinidad clearly illustrates the need for four-wheel drive tractors in the sugar cane industry.

RIGHT:
Roadless Ploughmaster 78 harvesting sweet potatoes in Barbados.

BELOW:
Roadless 120 with a Lambourn cab in 1976.

ABOVE:
Roadless's yard in Hounslow taken from Gunnersbury House in 1977. Compare this view with the 1964 photograph on page 99. The yard has not changed, but note how the surrounding area has become built up.

LEFT:
Roadless Ploughmaster 78 J Series equipped as an industrial towing tractor.

horse-power / hour.

In 1980, the 118 was matched against its nearest competitor in the Ford range, the 126 hp four-wheel drive TW10, and both tractors were using the same five-furrow Dowdeswell plough pulling 70 in. furrows. The Ford tractor, dearer by over £2,000, was also more powerful, but Roadless had a point to prove. Mervyn Ford with the 118 ploughed at over 5 mph, while the TW10 only managed 3.3. At the end of the day, the Ford had only turned in 2.4 acres per hour, while Roadless with 3.6 were in first position in the performance results for the sixth time.

The old Roadless 95 demonstrator, which had given such sterling service and came away with all the accolades at the 1972 and 1973 Long Sutton trials, had been used as a test-bed for many of the components for the new 118 by Roger Haynes, a contractor and farmer from near Evesham, who had bought his first Roadless tractor, a 4WD Super Major, in 1967. When looking for his second Roadless machine in 1970, his enquiries had brought him into contact with Ven Dodge and Mervyn Ford. He was eventually pointed in the direction of a 115, which he purchased from the Grantham Tractor Company. Impressed by his great enthusiasm for the make, Roadless offered him the opportunity to carry out some experimental trials for the company. Roadless preferred to use contractors for their development work as they knew all the components would be tested to the limit. Haynes eventually bought another 115, as well as the 95 test tractor which he fitted with a

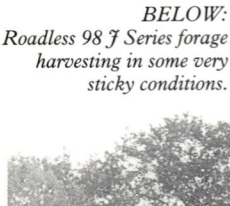

BELOW: Roadless 98 J Series forage harvesting in some very sticky conditions.

RIGHT: The Roadless 95 which was used as the development tractor for the 118, pulls a Bettinson direct drill while on trial with Roger Haynes.

144

2715E engine to bring it up to full 118 specification. All these tractors were used as test-beds for various new Roadless components, including axles and hubs.

During the late 1970s, Roadless sent a completely new axle to Roger Haynes for him to try out. The main feature of the new axle was a portal-type hub reduction with a single planet pinion permanently engaged with the internal ring gear. The planet in each hub was driven through a constant velocity, double universal joint which gave a 50 degree turning angle. All the axles previously used by Roadless had used swivel ball-type steering joints which only allowed a 28 degree angle of turn. The axle was fitted with a full differential-lock, and all the components were designed and built by Roadless, with the exception of the special universal joints which were bought in from Germany.

The idea behind the new portal axle was to compete with the products of the Schindler company whose Swiss-built axles had a similar steering angle, and were beginning to affect Roadless's sales. Ironically, the Schindler four-wheel drive conversion was marketed in the UK by Farm Tractor Drives of Ambaston, near Derby, a company started in 1977 by Arthur Battelle after leaving Roadless. The main advantage of the Schindler conversion for the Ford

TOP:
Roger Haynes's prototype 118 photographed in 1994.

ABOVE:
Roadless 118 J Series tractor.

LEFT:
Roadless 118 at the Long Sutton ploughing demonstration.

Roadless

*RIGHT:
Roadless 118 export model.*

tractor was the use of a side mounted transfer box, which allowed it to be easily fitted to new or used tractors without any modifications or affecting the use of the Ford Q-cab. Roadless had also experimented with a side-drive, but the unit was found to be too noisy, and was abandoned in favour of the trusted system of the transfer box sandwiched between the gearbox and rear transmission.

The portal axles went into production on two new four-cylinder Roadless models – the 780 and 980, based on the Ford 6600 and 7600 skid-units. The new tractors were given a high profile launch in July 1979 in a field at Fifield Farm in Bray, near Maidenhead, helped along by the provision of food and iced drinks set out on a trestle table in a marquee. With hydrostatic steering, and a turning circle of only 28 ft, the company claimed that the four-wheel drive tractors had the manoeuvrability of two-wheel drive machines. The assembled media, including the agricultural trade press, Fleet Street reporters, radio and representatives of BBC 1's *Tomorrow's World* programme, were given the chance to drive the tractors, and appeared impressed. Prices for the 780 and 980 were announced at £13,180 and £16,320 respectively with the 'K' Series cabs.

*ABOVE:
Roadless 120 J Series in Hertfordshire.*

*RIGHT:
Roadless tractors with the K Series cabs. Left to right are the 120, 118 and 98.*

146

ABOVE:
Roadless Ploughmaster 78 K Series at work in 1978.

LEFT:
Roadless 118 K Series mowing grass for silage.

Roadless

RIGHT:
Roadless 780 and 980 fitted with portal axles.

BELOW:
A Ploughmaster 78 and a 118 loaded on Roadless's Ford D Series lorry ready to leave Hounslow.

RIGHT:
Ploughmaster 78 and 118 on Roadless's stand at the Smithfield Show.

Unfortunately, the early track-record of the tractors was not as auspicious as the occasion of the launch, and some problems were experienced with the retaining circlip for the single planet pinion, which had a nasty habit of snapping when under load and falling into the ring gear. Modifications were made, but the tractors never really achieved the sales they deserved, and only just over a dozen of each were built.

It was becoming a time of changes at Roadless. Rupert Booth had retired in 1977 at the age of seventy-five. Vic Crockford, who had become technical director in 1973, took over as managing director, and a new chief engineer was appointed in the form of David Angier.

During the 1970s, the site at Hounslow had become a valuable commercial property. The town, conveniently situated between London and Heathrow, was on the Great West Road and served by both the Southern Region railway and the Piccadilly Line of the Underground. Naturally, it was becoming an important development area, and Gunnersbury House and yard, just off the main High Street, was in the centre of

148

Change and Decline

five 10,000 sq ft bays. Founded during the Second World War for aircraft production, the engineering firm had retained its links within the fields of aerospace and armaments, as well as offering facilities to the automotive, machine tool and processing industries. The company's main business was now the specialist manufacture of jigs, fixtures and production-line tools and machinery. Its impressive list of customers included Rolls-Royce, Hawker Siddely, the Bank of England and the Ministry of Defence, to name a few. Originally, Walter Lawrence was only interested in selling the business and it. Concerned about the prospect of a compulsory purchase on the site, Roadless had agreed a sale with County and District Properties back in 1973, but several years of delays had seen the completion date for the purchase moved back to July 1979.

The company now had the capital investment to purchase new premises, but both Rupert Booth and Vic Crockford had been looking for a new long-term site for several years without success. A chance response to an advertisement in the *Financial Times* offering a business in a modern factory led Roadless to Walter Lawrence Engineering at Sawbridgeworth, just north of Harlow in Hertfordshire. A subsidiary of the Walter Lawrence construction group, the company was operating from a recently completed and purpose-built factory unit with

LEFT:
This photograph clearly shows the 50 degree turning angle of the portal axle.

BELOW:
Close-up of the portal axle hub showing the double universal joint drive to the single planet gear.

LEFT:
Roadless 980 at work with a five-furrow Dowdeswell plough.

RIGHT:
Cultivating with a Roadless 980.

BELOW:
The premises of RTL Engineering at Sawbridgeworth.

Change and Decline

*LEFT:
Roadless's old premises at Hounslow just before demolition. The road in the foreground is already a new addition built straight through the middle of the old yard.*

Roadless only wanted to purchase the factory. After much negotiation, a deal was finally struck, and Walter Lawrence Engineering was purchased lock, stock and barrel in April 1979, with Roadless acquiring both the company and the premises.

Roadless now had both its new site, and an engineering base from which it could expand into new fields with new products. The company could see no future in total reliance on the four-wheel drive tractor market as sales of Roadless tractors had begun to decline. In some ways, the

*LEFT:
Roadless Ploughmaster 78 with a Duncan Supercab at Sawbridgeworth.*

ABOVE: Roadless 98 at Sawbridgeworth. The Lambourn cab on this tractor has been fitted with air-conditioning.

RIGHT: The prototype Roadless hydrostatic forestry tractor built for the Forestry Commission in 1969.

company was a victim of its own success. During the mid-1970s, when there had been a nearly two-year waiting list for Roadless tractors, many farmers had looked elsewhere and bought cheaper continental and east European four-wheel drive tractors, such as Same, Belarus and Zetor, which had then gained a toe-hold on the UK market. Roadless was also competing against the Schindler conversions, and the Ford Motor Company, using Carraro or ZF axles, had four-wheel drive versions of most of its tractors either out or in the pipeline, and most major manufacturers were following suit. Roadless had proved the efficiency of the four-wheel drive tractor, and now everyone wanted to build one.

The move from Hounslow was not finally made until December. An extension had been agreed, and Roadless paid six months rent on the old property. A new unit was opened up at Browells Lane in nearby Feltham to assemble the tractors from components made at Sawbridgeworth, and to provide work for the employees from Hounslow, several of whom had been with the company for thirty or forty years.

Roadless was restructured in July 1980. The old Roadless Traction Limited became Roadless Holdings Limited, with two subsidiaries. The first of these was a new Roadless Traction Limited to deal with assembly, distribution, sales and promotion of the tractors and associated machinery. The employees of the former Roadless Traction Limited carried on in the new company without a break in employment. The other subsidiary was RTL Engineering, formed to carry on the business of Walter Lawrence Engineering at Sawbridgeworth. Most of the

Change and Decline

ABOVE:
The prototype Roadless hydrostatic skidder on trial with the Forestry Commission.

LEFT:
The Mark III prototype hydrostatic skidder built by the Forestry Commission in 1973.

ABOVE:
The Mark III prototype hydrostatic skidder on trials at Alice Holt Forest in Hampshire.

RIGHT:
The Mark III prototype hydrostatic skidder was powered by a 63 bhp International engine.

Angier became technical director.

An alternative product and one to benefit from the new engineering capacity at Sawbridgeworth was the Roadless Logmaster. Although not announced to the public until 1980, the story behind this innovative forestry skidder went back twelve years. In 1968, Roadless was approached to design a timber forwarder for Till Hill Forestry to use in its various holdings of standing timber. Vic Crockford drew up plans for a machine based on a Ford 3000 with a three-speed hydraulically driven rear axle powered by a pto pump. At Till Hill's request, the design was submitted to the Forestry Commission for a second opinion. Unfortunately, the report from Rodney Ross, the Commission's chief research and development engineer, was not favourable, and so the forwarder was not built. A similar machine, based on a small Massey Ferguson tractor and fitted with mechanical drive, did, however, undergo extensive trials with the Commission's Work Study Branch at a later date.

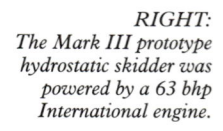

work of the engineering company continued as before under Roadless's ownership, and assembly fixtures were even supplied to the Delorean sports car company in Ireland. Vic Crockford became managing director of all three companies. Ven Dodge was made sales director of the new Roadless Traction Limited, and David

Roadless had enjoyed a long association with the Forestry Commission, and been involved in many joint ventures. Although the timber forwarder venture had not amounted to anything, discussions surrounding the project led to the company being invited by Rodney Ross to a demonstration of a Japanese Mitsubishi forestry tractor. The demonstration, held in Hafren Forest in Wales, was attended by Rupert Booth, Vic Crockford, Charles Skelton and Ven Dodge. After the demonstration, Ross asked Vic Crockford for his opinion of the machine. As Crockford was somewhat critical of the suitability of the Mitsubishi for use in all forestry conditions, Roadless was invited to submit a design for a machine to do the job better. An agreement was reached whereby Roadless would pay for the costs of design and drawing work on the understanding that the Forestry Commission would purchase the first prototype.

Roadless had to produce a high specification machine that would satisfy certain important criteria laid down by the Forestry Commission.

ABOVE:
The Mark III skidder demonstrates the degree of articulation that can be achieved by this machine.

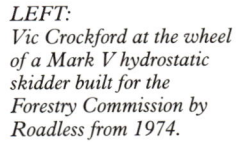

LEFT:
Vic Crockford at the wheel of a Mark V hydrostatic skidder built for the Forestry Commission by Roadless from 1974.

155

RIGHT:
Mark VII hydrostatic forestry skidder based on a Ford 4000 skid-unit.

RIGHT:
This photograph illustrates the ground clearance of the skidder and the clean lines of the underpan designed so as not to snag on the undergrowth.

By the late 1960s, man-made forests planted thirty years previously were coming to maturity. The trees, planted in lines, required thinning by taking out one in three trees, and soft-wood thinnings needed removing after only twenty to twenty-five years. The plantations had been designed so that the thinnings could be extracted by a machine, known as a skidder. Although the Forestry Commission managed half the total UK forest acreage, much of it was on poorer land such as the soft peaty areas around the Scottish Borders. Here the trees were planted on ridges, often on ground which was previously ploughed or crisscrossed with drainage channels.

Any machine capable of coping with this terrain had to be an exceptional vehicle. It had to have sufficient ground clearance to operate over stumps and logs, but also have the stability and low centre of gravity needed to work on steep slopes. It had to be able to negotiate deep furrows, drainage channels and ditches, and yet be highly manoeuvrable and narrow enough to work on the tracks between the trees. The weight of the machine also had to be kept to a minimum so as to not damage the root systems of the standing trees.

After much consultation with the Forestry Commission, Vic Crockford completed the drawings for the skidder in February 1969, and the prototype was ready by May. The four-wheel drive machine had an articulated hollow-steel chassis to allow the wheels to stay in contact with the ground on rough terrain. Pivot-steering provided excellent manoeuvrability and gave the skidder the ability to 'walk' itself out of difficult patches. The front chassis also incorporated a safety frame for operator protection.

To withstand the excessive demands of the arduous conditions under which it was to work, a hydrostatic drive was chosen to protect the transmission from overload. This also gave the machine the advantage of infinitely variable speeds, and incorporated a dynamic braking system through a counter-balance valve. A 75 hp Ford 5000 engine drove two in-line dual capacity Carron pumps powering Carron-Becander wheel motors, supplied by Carron Langdale Limited of Falkirk. As the trees were on the small side, two hydraulic winches and fairleads suitable for timber skidding were fitted, as was a front-mounted hydraulic blade for log rolling.

The prototype Roadless skidder was shown at the Forestry Commission's Jubilee Exhibition

ABOVE:
The joy-stick control for the skidder which was preferred by the Forestry Commission.

BELOW:
The operator of the hydrostatic skidder pays out the hydraulic winches by remote control from the radio pack on his chest.

ABOVE:
The Roadless Logmaster at Sawbridgeworth.

RIGHT:
Roadless Logmaster in use with the Forestry Commission. This photograph clearly illustrates the type of thinnings the machine was designed to handle.

supervision of Rodney Ross. The first machines, the Mark I and II, were completed in 1971. Designed for first thinnings, they were similar in concept to the Roadless skidder, but were smaller and lighter, and powered by 43 bhp International 444 skid-units. The Mark III followed in 1973. This larger machine was designed for second thinnings and logging. Power came from an International 63 bhp D239 engine, as used on the 574 tractor, driving a single over-centre Sundstrand pump. Oil from the single pump was fed through three flow-dividers to the four Carron wheel motors. Problems were encountered with too much pressure in the system, and a Mark IV was brought out with the same engine driving twin Sundstrand pumps feeding the wheel motors through a Brevinni flow-splitter. This system seemed to work best, so a Mark V was designed using a similar hydraulic arrangement, and a 62 bhp Ford 4000 engine.

at the Bush Estate in June, before going to the experimental station at Alice Holt Forest in Hampshire. Initial trials were encouraging, but revealed a few defects, including problems with oil leaking past the piston in the wheel motors when hot. The Commission also felt it wanted something less heavy, and set about building its own prototypes.

The Forestry Commission prototype forestry tractors were built at Alice Holt under the

After thoroughly testing the prototypes in Scotland, Wales, the Kielder Forest in

LEFT:
The Roadless Logmaster demonstrates its ability to cope with forest terrain.

Northumberland and the Forest of Dean in Gloucester, the Forestry Commission asked Roadless to build a batch of twelve machines for delivery from June 1974, at a provisional cost of £8,500 to £9,000 each. Production was initially delayed due to shortages of engines and steel, but the order was eventually completed by March 1977, incorporating various modifications which resulted in Mark IV to Mark IX machines. Some of the skidders built for the Commission had a joy-stick in place of the steering wheel, to control forwards and backwards movement,

LEFT:
A rear view of the Roadless Logmaster showing the hydraulic winches and fairleads.

ABOVE:
Roadless Logmaster supplied to Cumbrian forestry contractors, Sheffield and Company Limited of Southwaite near Carlisle in 1979.

RIGHT:
Sheffield's Logmaster extracting timber from 'windblown' areas of Newcastleton Forest. The root bowls of each tree left deep pits making the job difficult. Note how the operator can winch the trees in by radio control while standing safely to one side.

steering, braking and differential lock. Emergency braking on all the machines was effected by lowering the rear section of the machine by hydraulic rams so that the mudguards gripped the rear tyres.

One of the most revolutionary aspects of the skidders was the winch arrangement. The winch could be operated by radio control from a pack strapped to the driver's chest. After paying out the fairlead, he could fasten the scattered felled timbers with sliding chokers that clipped over the wire rope, and then stand safely to one side as he instructed the winch to pull in. The ends of the logs were then gathered in together in a bunch and suspended from the rear of the machine ready to be hauled to the road.

While manufacturing the skidders for the Forestry Commission, Roadless could see many aspects of the machines that could be improved or refined, and so Vic Crockford and David Angier, who had considerable experience in hydraulics, started designing a more advanced forestry tractor. The result was the Logmaster, which retained the same basic concept, but incorporated many design improvements over the earlier machines. The Ford power-unit was found to set up too many vibrations through the machine, so the Logmaster was powered by a four-cylinder Perkins 4.236 rated at 76 bhp. A variable-stroke hydraulic pump, fed through flow-dividers, drove four dual-capacity Poclain wheel motors, which gave two speed-bands, one for forest work and one for travelling on the road.

Roadless did not like the joy-stick arrangement, and used a steering wheel controlling a dual-sensitivity hydrostatic steering system, which gave one-and-a-half turns lock-to-lock for forest work, and four turns for road work. The winch system was refined so that the fairleads would payout automatically under tension, and operator comfort and safety was not forgotten by the provision of an anti-vibration mounted quiet cab with a Bostrom seat. Front visibility was improved by the adoption of a sloping bonnet.

At £29,950, the Logmaster was not cheap. However, a standard agricultural four-wheel drive tractor when equipped for forestry work would cost nearly £24,000, and the company proved the Logmaster could easily match the output of two tractors working flat-out. It was also capable of working on slopes where the extraction of timber had previously only been possible by skyline winch. The Forestry Commission placed an order for thirty-six machines, and the first machine, serial number 001, was delivered in May 1979. Another was sold in July to the Cumbrian forestry contractors, Sheffield and Company Limited of Southwaite, near Carlisle, who were very impressed with its performance. Roadless were slightly disappointed to find out that the Forestry Commission had also bought some new County tractors, but were somewhat compensated by successfully tendering to fit them with winches, roller-blades, pan and brush guards, all fabricated at Sawbridgeworth.

Roadless exhibited the skidder at various shows and demonstrations throughout 1980 and 1981. At an Institute of Professional Foresters exhibition near Penrith in Cumbria, Mervyn Ford demonstrated the versatility of the transmission by driving half-way down a one-in-one slope, bringing the machine to a halt, and then reversing back up again – a feat for which he was awarded by spontaneous applause from the crowd of onlookers. As the first British designed and manufactured forestry skidder, the Logmaster met with acclaim wherever it was shown, and attracted the company plenty of good publicity. Roadless rightly felt it had a winning product that would see it into the 1980s and compensate for the falling four-wheel drive sales.

Unfortunately, the country was in a recession,

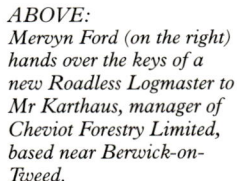

ABOVE:
Mervyn Ford (on the right) hands over the keys of a new Roadless Logmaster to Mr Karthaus, manager of Cheviot Forestry Limited, based near Berwick-on-Tweed.

LEFT:
The Logmaster on Roadless's stand at the 1981 Royal Show.

ABOVE:
The Roadless Logmaster is tested for stability.

RIGHT:
County 774 tractors, modified for forestry use and fitted with winches, roller-blades and brush guards by Roadless at Sawbridgeworth for the Forestry Commission.

RIGHT:
One of the last Roadless Logmasters built at Sawbridgeworth.

a private contractor at Berwick-on-Tweed. Interest was shown from Denmark and New Zealand, but the company was still faced with unloading four expensive machines at heavily discounted prices on to a depressed market. This was a serious blow to Roadless who had invested a lot of money into the Logmaster project.

The importance of the Logmaster cannot be stressed to highly. It was believed to be the world's first hydrostatic forestry skidder, and its design was a remarkable and imaginative achievement for such a small concern as Roadless. Users of the Logmaster could not speak and the slump was affecting the pulp processing and paper industry. During 1980, important pulpmills were closed at Fort William and Ellesmere Port, and the demand for soft wood fell overnight. A further blow came when the Government ordered the Forestry Commission to move out of harvesting operations, and sell the trees as standing timber to private contractors. The Commission's order for the Logmaster was cut to twenty-four. Roadless had already built thirty, and had the materials on hand for the other six. One machine was sold to the Irish Forestry Commission, and another went to

too highly of its performance, and second-hand machines are still in demand today. The company was justly proud of the Logmaster, not only had it bravely broken in to a new market, it had also led the field in innovation and design. Sadly, unfortunate circumstances meant that Roadless never received the financial benefit from the project that it deserved.

The company still had other projects on the boil, including the development of the Amex Workhorse. This basic 'no frills' mini-tractor was the brainchild of an independent automotive designer, Valerian Dare-Bryan, and was aimed at the third world countries requiring a simple, low maintenance machine. It was powered by a rear-mounted and air-cooled Ducati twin-cylinder diesel engine delivering 22 hp. The four-speed gearbox and rear transaxle were derived from Mini-Metro components, and the drive from the engine was via a flat belt, tensioned by a jockey-wheel acting as the clutch. Fitted with a very basic category 1 hydraulic lift, it was envisaged that the machine could also appeal to smallholders and stockmen. In practice, the tractor was found to be unreliable. Mervyn Ford took the prototype home to test it by ploughing his garden, and the drive belt burnt out in the first hour. Some interest was shown from Zimbabwe, and David Angier re-designed the machine to use dumper components, but none were built and the project was shelved.

The company experienced more success with the Roadless Teleshift. This twin telescopic boom materials handler was designed by Ron and Tony Collins of RWC (Developments) Limited of Ledbury, Herefordshire. The Collins brothers had originally been instrumental in the development of the single central boom system as used by Sanderson on the Teleporter. The twin boom arrangement made the machine more compact, and enabled the cab to be centrally mounted for better visibility. Ease of maintenance was also considered, and simplified hydraulic systems were used. The machine, designated the Teleshift TS230, was based on an International

TOP:
The Roadless Amex Workhorse designed for use in third world countries.

ABOVE:
Mervyn Ford at the wheel of the Roadless Amex.

LEFT:
Roadless Amex at work with a single furrow plough manufactured by Lewis Agri-Equipment of Evesham.

ABOVE:
Roadless Teleshift TS230.

RIGHT:
The Teleshift TS230 working in a silage clamp.

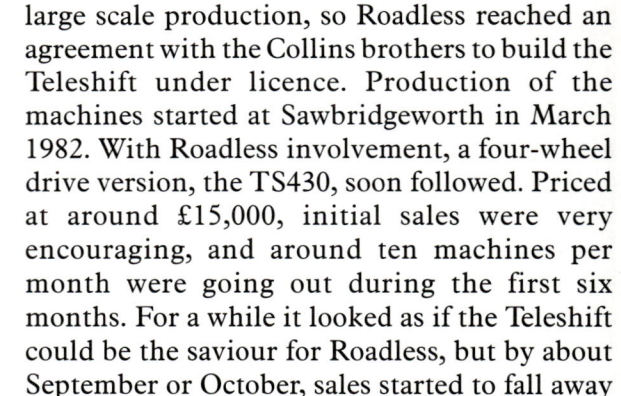

RIGHT:
The four-wheel drive Roadless Teleshift TS430.

584 skid-unit fitted with an eight-speed shuttle gearbox.

RWC did not have the facilities to expand into large scale production, so Roadless reached an agreement with the Collins brothers to build the Teleshift under licence. Production of the machines started at Sawbridgeworth in March 1982. With Roadless involvement, a four-wheel drive version, the TS430, soon followed. Priced at around £15,000, initial sales were very encouraging, and around ten machines per month were going out during the first six months. For a while it looked as if the Teleshift could be the saviour for Roadless, but by about September or October, sales started to fall away and Roadless was running into difficulties.

Financial pressure had led to the closure of the Feltham depot in 1981. Four-wheel drive tractor production continued at Sawbridgeworth, but was only limited as price competition from Ford had made the market very tight. Only ten complete tractors and ten axle kits went out in 1982. Comparing this to the figure of nearly 300 tractors that were produced at Hounslow in 1975 indicates how dramatically the fortunes of a company can

Change and Decline

LEFT:
The Roadless 6/2, a two-wheel drive version of the Roadless 118 built in March 1982, with a Hallam road grader.

BELOW:
The Roadless 6/2 seen fitted with a Duncan Supercab.

change over seven short years.

Roadless bravely tried to encourage sales by bringing out new models. The S Series Roadless tractors were launched, based on the new Ford 10 Series range. The Ploughmaster 78 now had the benefit of 86 bhp, and the Roadless 98 was rated at 103 bhp. Two new equal-size wheel models were announced, based on the 780 and 980, with the portal axle inverted and different ratios in the transfer box. Only one, a 9804, was built. It was shown at Smithfield in 1981, but was converted back into a 980 after the show. A two-wheel drive version of the 118 was offered in March 1982 as the Roadless Six-2. Powered by the 2715E engine, which now developed 125 bhp, it was available either with a Duncan Supercab, or without a cab for export. One tractor was built, and not sold was converted to four-wheel

LEFT:
Roadless 118 S Series.

165

Roadless

*RIGHT:
Roadless 9804 built for the 1981 Smithfield Show.*

*RIGHT:
The Roadless 9804 used an inverted version of the portal axle to give an equal-size four-wheel drive arrangement.*

Change and Decline

LEFT:
Roadless 118 S Series at Sawbridgeworth in 1982.

drive in August.

RTL Engineering was now having to rely on outside work to keep it going. Unfortunately, the recession was hitting both the automotive and the aerospace industries, and business had slackened off considerably. Without the cushion of work from Roadless Traction, the company was no longer viable and went into voluntary liquidation in December 1982.

Roadless Traction Limited struggled on for a time until lack of business also forced it into voluntary liquidation in March 1983. The last tractor to be sold by Roadless was an S Series 120 which had been delivered to British Telecom in February. This machine was finished in a striking blue and yellow colour scheme, and used the new industrial Ford 2725 engine developing 130 hp. Two tractors remained in stock after the company went into liquidation, and were disposed of by the liquidators. A Roadless 120 was sold to a farmer in April, and a Ploughmaster 78 was purchased by the House of Goodness, a religious cult from Oxfordshire, for the knock-down price of £7,500 in June.

Sixty-four years of pioneering design and innovation had come to an end, as a combined mixture of uncontrollable and unforeseen circumstances brought this proud company to its knees. The parent company, Roadless Holdings, which owned the land and premises, remained in existence in the vain hope that a quick sale of the Sawbridgeworth site would leave it enough capital to buy into a new business. However, it was not to be as the recession bit deeper, and hopes faded as no buyer could be found for the premises.

Roadless had tried everything in its power to stay afloat. The company had forseen the diminishing market for its four-wheel drive tractors, but felt that with the Logmaster it had armed itself with a new product to combat falling sales. Sadly, the Logmaster had hit the market at the wrong time, and the success it warranted never came. The feelings at Roadless at the time of the liquidation are best summed up by the words of Ven Dodge, 'We went down with flags flying and full honour'.

BELOW:
The last tractor to be sold by Roadless, a 120 fitted with a Ford 2725 engine and supplied to British Telecom in February 1983.

EPILOGUE
1983 to 1996

Following the liquidation of Roadless in 1983, the company's dealer network was taken over by RWC, and Ven Dodge, Mervyn Ford and David Angier joined the Collins brothers to develop and promote the Teleshift. The Benson Group, who already owned County Tractors, bought RWC in 1985, and the Teleshift was sold in a four-wheel drive version as the 404 with a Ford engine. Benson, in turn, sold the Teleshift to Matbro in 1991, and it is still in limited production today at Tetbury in Gloucestershire.

Vic Crockford returned to engineering, but is now retired. Ven Dodge retired six months after moving to RWC, and now devotes his time to amateur and freelance photography. Mervyn Ford, today, runs his own business, Axis Plant Hire, in Warwick.

The business of Roadless Traction, including the manufacturing rights to the four-wheel drive tractors and the Logmaster, was sold to the West Country Ford dealers, L. F. Jewell Limited, together with the spares and stock-in-hand which were transferred to Somerset in June 1983. A move which took five days and required fourteen 40 ft articulated lorries to complete. Jewells, founded in 1960 by Leslie F. Jewell, the son of a Somerset farming family, operated from its main depot, known as the Market, in Bath Road, Bridgwater. Graham Melvin, who had joined RTL Engineering in charge of stock control just over a year previously, was taken on by Jewells, together with a fitter from Sawbridgeworth, to run the new operation from premises in Wyld's Road, Bridgwater, on an industrial estate about half a mile from the Market.

The first few weeks were very arduous for Melvin as he tried to sort out the spares and learn the intricacies of manufacture, assembly and parts identification. Over the next few months, Melvin gradually picked up the threads of the Roadless business, not least in the sale of parts overseas – resulting in an encouraging order from St Kitts for spares worth over £60,000 placed in 1984.

Although parts and axle sales were the mainstay of Graham Melvin's Roadless division at Jewells, a few complete tractors

BELOW: The Jewelltrac 120 and the Jewelltrac 103 at the Market in Bridgwater.

RIGHT: The Jewelltrac 120. This tractor is believed to have been sold to Bruff for drainage work.

LEFT:
The Logtrac – Jewells' version of the Roadless Logmaster.

BELOW:
The Logtrac photographed at Jewells' Cannington depot.

were also produced and sold as Jewelltracs. The first of these, the Jewelltrac 103, was built at the Market under the supervision of the branch manager, Steve Kitch. The 103, based on the Ford 7610 skid-unit, was essentially the same tractor as the Roadless 9804 using the inverted portal axle that had remained in the stock from Sawbridgeworth. Only one was built. Jewells also inherited a couple of unfinished machines, including a Roadless 120 which was completed and sold as the Jewelltrac 120 without a cab, and is believed to have gone to Bruff for drainage work.

Graham Melvin's division also built twenty-four equal-size four-wheel drive tractors based on the 116 bhp Ford 8210 skid-unit. Designated Jewelltrac 116, these machines were finished to an industrial specification, and the first batch of five went to the Wimpey construction group in 1984. One of these tractors is thought to be still at work in South London as a platform for a mounted crane. The other nineteen Jewelltrac 116s were completed between December 1985 and March 1986, and were built as industrial shunters to the order of the Tanzanian Government for use on the state railways. Six of these machines were later purchased by KLM, the Dutch airlines, and shipped back to Holland.

One unfinished Logmaster had remained in the stock purchased from Sawbridgeworth. This was completely re-designed by Jewells – much of the old pipework was dispensed with and the engine was turbocharged for more power. Known as the Logtrac, it underwent evaluation trials on the Fortescue Estate, near South Molton in Devon, before being sold to a Mr Pierce Miller of Dalrymple, near Ayr in Scotland, in 1991.

An important part of Graham Melvin's division's business was the sale of axles to Phoenix Engineering Company Limited of Chard, also in Somerset. Phoenix, specialist manufacturers of road gritters and asphalt

RIGHT: One of Roger Haynes's two 115 tractors photographed at the Agrosave yard near Evesham.

pavers, had used Roadless axles and transfer boxes for many years on its machines. When Jewells took over the Roadless business, it was natural that with the two companies now so close, the association would continue and grow. Jewells supplied a great number of specially designed four-wheel drive axle kits, consisting of the axle, transfer box, track-rods, torque-control clutch and special gears to fit the Ford 8210 skid-unit used by Phoenix. The Phoenix kits were sold to the company for between £5,000 and £8,000 each, and were good business for Jewells.

In 1985, Jewells was bought by the Kellands group, a company which held major industrial and construction machinery franchises across Devon and Somerset. Graham Melvin's Roadless division was moved to the Market in Bridgwater in about 1987, and relocated again two years later to Jewells' Cannington depot. Unfortunately, the Kellands group ran into financial difficulties, and both it and Jewells went into liquidation in November 1991.

The Roadless division was sold to Roger Haynes, the farmer and contractor from Evesham who had carried out much of the test work for Roadless during the 1970s and 1980s. The stock and spares were moved from Bridgwater to premises known as the Sidings, at Harvington near Evesham. For Roger Haynes, the acquisition of the Roadless business, which he operates under the name of Agrosave is the culmination of a lifetime's enthusiasm for the make. Today, Agrosave supplies parts for Roadless tractors to wherever they may be working all over the world, and the company has recently responded to the demand for Selene spares which it now also stocks.

As a footnote, the name of Roadless Traction was revived in 1994 by Robert Eyre, an agricultural engineer from near Alford in Lincolnshire, to promote his new designs of low ground pressure agricultural tractors on rubber tracks.

Appendices

APPENDIX 1
Selene and Four-Wheel Traction

RIGHT:
One of Selene's four-wheel drive conversions of a Fordson E27N Major fitted with a Perkins L4 diesel engine.

BELOW:
A 1954 Fordson Diesel Major fitted with Selene four-wheel drive photographed in Ireland.

Although both were independent companies, Roadless and Selene had a long association, and many of their developments ran along similar lines. Selene was an important player in the pioneering of four-wheel drive tractors, and a brief history of the company provides interesting parallels to the Roadless story.

No new tractors were being manufactured in Italy in the immediate postwar period, leading to a lack of machines for the home market. Selene was formed by two associates, Dr Sion Segre-Amar and Signor Cantone, to exploit this shortage and import used tractors from Britain. The company was based in the town of Nichelino, near Turin, and the name, Selene, is believed to have been derived from the name of Segre-Amar's wife.

Cantone later left the company and went to farm in the Mato Grosso area of Brazil. The business grew under Segre-Amar, and by 1950, he was importing mainly Fordson E27N Major and Ferguson TE20 tractors, which were overhauled, and often fitted with Perkins diesel engines before being sold on the Italian market. He also had the Italian distribution rights for the Perkins engines through an associate company, Riberi.

Many of the tractors, including most of the Fergusons, were supplied to Selene by Robert Eden and Company. Based in London, the Robert Eden company was one of the most prominent exporters of used farm tractors in Britain. Founded in 1948, the company was joined two years later by William Fuller, who took over the business in 1952. Incidentally, Fuller was responsible for the Red Machinery

Appendices

LEFT:
A Ferguson FE35 fitted with the Selene four-wheel drive conversion.

BELOW:
Selene conversion of a 1958 Massey Ferguson 65 Mark 1 using the pto transfer box drive.

BOTTOM:
Close up of the Selene ground-speed pto drive transfer box fitted to the Massey Ferguson 65.

Guide, a publication which he started in 1959. When in England, Segre-Amar would stay at the Grosvenor Hotel, not far from Eden's offices in North Audley Street, where he would be met by Fuller to sort out the purchase of the next batch of tractors for export. Italian import regulations restricted the import of complete tractors, and so the machines were dismantled by the London company, and exported as spares. On arriving in Italy, the Fergusons, normally petrol-engined Continental models, were reassembled and fitted with Perkins P3 engines.

However, by the beginning of the 1950s, new tractors were rolling out of the factories of the main Italian manufacturers, including Fiat, Landini and Same, and Selene, hampered by expensive import licences, found its sales badly affected. Another blow came when Ford brought out the improved new E1A Major in late 1951, and the Italian company found itself stuck with several old model E27Ns which were not selling.

Selene looked for a new niche in the market, and discovered that few four-wheel drive tractors were available either in Italy, or any other country of the world. The company set about building a conversion to fit the Fordson, initially only to encourage sales of the old models. Ex-US Army war surplus axles were bought in from various suppliers, and the drive was provided by a transfer box designed to fit between the tractor gearbox and rear transmission. Selene was not the only company to use military vehicle axles in Italy; Ravasini built a couple of crab-steer four-wheel drive machines fitted with Chevrolet

173

Roadless

RIGHT:
A Massey Ferguson 65 fitted with the Selene four-wheel drive conversion exhibited at the 1959 Essex and Suffolk Show by Robert Eden and Company.

BELOW:
Massey Ferguson 135 with a Selene four-wheel drive conversion driven by a pto transfer box.

BOTTOM:
Robert Eden pto drive transfer box.

axles front and rear during the mid-1950s.

Initially, Selene built sixteen four-wheel drive tractors, based on the old E27N Majors and fitted with Perkins engines. Ten were sent out to Signor Cantone in Brazil for use in ricefields, and six were sold in Italy. Encouraged by their performance, and having gained approval from the Ford Motor Company, Segre-Amar patented the conversion in Italy in December 1953, and commenced building four-wheel drive versions of the new E1A Major. The trade name for the conversion, Manuel, was derived from the name of Segre-Amar's son, Emmanuel.

The story of the licensing agreement between Selene and Roadless for the four-wheel drive Fordson conversion to be built in the UK has already been documented earlier in the book. However, the Fordson was only one conversion among many designed by the Italian firm. In about 1953, Selene also brought out a four-wheel drive conversion for the Ferguson tractor, using a sandwich transfer box driving a Jeep front axle. The tractor was fitted with a Perkins P3 engine and was designed for hill-side ploughing in the Po Valley area of Italy. Later examples were used in ricefields and and several went out to Puerto Rico. Selene went on to patent as many different systems of four-wheel drive as possible, including a system for Fergusons using a transfer box driven off the rear ground-speed pto drive. Eventually the company also built four-wheel drive versions of Fiat, Landini, Renault, Deutz, Steyr, Nuffield and David Brown tractors, using several different transfer box systems and army-surplus axles. Most of the four-wheel drive systems were conceived by Selene's chief engineer, Signor Torchio.

A few of the four-wheel drive Ferguson tractors with the ground-speed pto transfer box were imported and sold in Britain by the Robert Eden company, who later distributed the 4WD kit in the UK under licence. A Ferguson FE 35 with the four-wheel drive conversion was exhibited by the company at the 1956 Bath Royal Show. Four-wheel drive versions of the Massey Ferguson 35 and 65 tractors followed, and after the new MF range was introduced at the Smithfield Show in 1964, Eden offered Selene conversions for the 135, 165 and 175 tractors, still using the ground-speed pto drive. William Fuller was not entirely happy with the strength of the Selene pto transfer box and set about designing his own improved version for the Massey Ferguson tractor range. This box, distinguished by the letters RE stamped on the casting, was sold with the Selene axle as the Eden Manuel conversion from 1965.

The ground-speed pto drive had its limitations, the main drawback with the system being that the pto was no longer available to drive any attached implements or machinery. In 1968, Fuller reached an agreement with Selene allowing him to use a sandwich transfer box of his own design. Manufacture of the new conversion was undertaken by a separate company set up by Fuller, and known as Four Wheel Traction Limited, also operating from North Audley Street. New axles were sourced using British-built commercial vehicle components, and a new range of four-wheel drive Massey Ferguson tractors was launched at the 1969 Royal Show. The company also exhibited a Massey Ferguson 165 featuring lift-up front hub reduction boxes to give an equal-size wheel configuration of four-wheel drive.

In 1971, Four Wheel Traction designed four-wheel drive conversions for the Leyland 255 and 270 tractors. Some input in the design work was received from Bray who had ceased to build its own four-wheel drive version of the Leyland tractor. A drop box mounted on the underside of the tractor gearbox contained a gear train which drove a centrally mounted propellor shaft, a similar arrangement to that used on the earlier Bray Nuffield conversions. Two prototypes were built, and sent to work on farms in Sussex. The technical department at Leyland was willing to make the necessary alterations to the tractor gearboxes to accommodate the conversion, but the sales and marketing division were not keen on the idea, and no more than the two examples were produced.

Meanwhile, Selene had built conversions for

TOP:
Selene axle fitted to a Massey Ferguson 165 using the Eden pto transfer box.

ABOVE:
Four Wheel Traction Massey Ferguson 175 on trial in March 1968.

RIGHT:
Four Wheel Traction's stand at the 1969 Royal Show.

BELOW:
Equal-size four-wheel drive conversion of the Massey Ferguson 165 made by Four Wheel Traction.

RIGHT:
Four Wheel Traction 29:1 Superslow reduction box for Massey Ferguson tractors.

arrangement obviously had some influence on Leonard Tripp's early designs for the Roadless 115. A 4RU version of the Massey Ferguson was also available based on the 175 and 178 skid-units.

Selene four-wheel drive conversions became very popular in France, Holland, Belgium and all four tractors in the new Ford 6X range, from the 2000 to the 5000, introduced in 1964. The Italian company's own planetary axle for the Ford 5000 was launched at the Verona Show in 1971 and marketed as the 5000R conversion kit. The Ford 7000 became available with the Selene axle later the same year.

Selene also made an equal-size four-wheel drive version of the 5000. Known as the 4RU, the tractor was extended by the use of spacer boxes, and the engine was mounted forward of the front axle. The driver's seat was also moved forward to enable cranes, winches and other attachments to be fitted over the rear axle. The

Norway, where they were often factory fitted and marketed through the Ford tractor dealerships. Towards the end of the 1970s, Selene had out-grown its Nichelino works and needed to expand. Finding it difficult to get permission to extend his factory, Segre-Amar sold out to the Swiss Schindler company. Under Schindler's ownership, the Italian firm was slowly wound down and eventually closed. Emmanuel Segre-Amar went to Switzerland to work for Schindler, which also moved out of the four-wheel drive market in about 1985 to concentrate on escalator, lift and railway stock manufacture. A small company, known as Crupi, is thought to still exist in Italy at Lovero Valtellina, Sondrio, supplying Selene spares.

Four Wheel Traction no longer manufactured 4WD conversions after Massey Ferguson introduced its own range of four-wheel drive tractors in the late 1970s. Apart from the four-wheel drive conversions, the company had also developed a 4:1 reduction gearbox for use on Massey Ferguson tractors. This was joined in 1970 by the Superslow 29:1 creep box. For a time, the company had also marketed a turbocharger kit to fit the four-cylinder MF tractors. William Fuller had also experimented with fitting power steering to a four-wheel drive Massey Ferguson 135 for use on his own farm. This led to the introduction of power steering kits to suit both the Massey Ferguson and Ford tractor ranges. Although not offered in the kits produced for sale, Fuller's own tractor was also fitted with an extra ram to slew the axle and decrease the turning circle, similar in idea to the Super-steer axle introduced by New Holland Ford on its Series 70 tractors in 1994. Today, Four Wheel Traction still manufactures power steering kits for the Ford 6600, 5600, 5000, 4600, 4000, 3600 and 3000 tractors, and reduction gearboxes to fit several Massey Ferguson models.

ABOVE:
Selene 4RU equal-size four-wheel drive conversion of the Massey Ferguson 175 tractor.

LEFT:
Selene MF 175 4RU photographed in Italy.

BELOW:
Selene four-wheel drive axle fitted to a Dutch Ford 7000.

177

APPENDIX 2
J. J. Thomas Tractors

During the 1970s, Roadless could see a need for a six-cylinder two-wheel drive version of the Ford tractors, both for export and the home market where more power was required, such as for forage harvesting operations. Other firms specialising in Ford conversions were thinking along similar lines, and Ernest Doe and Sons Limited of Essex had built the Doe 5100, which was based on the Ford 5000 fitted with a six-cylinder 2704E engine, using sub-frame castings bought in from EVA of Belgium. Unfortunately, the machine attracted adverse reaction from the Ford Motor Company who would not approve six-cylinder two-wheel drive versions of its tractors, and only three or four were built.

The Ford company, while happy to supply skid-units for four-wheel drive conversions, felt that any two-wheel drive machines produced would affect the sales of its own tractors, especially of the new 7000. Roadless had built six-cylinder two-wheel drive Super Major tractors for export, and had experimented with similar conversions on the 5000, and at least one, designated Six-2, had gone to Saudi Arabia. But the company felt it should not actively pursue the manufacture of two-wheel drive units and upset the arrangements for the supply of Ford skid-units.

However, one man who managed to produce a considerable number of six-cylinder conversions was J. J. Thomas, who got round the problem of getting skid-units from Ford by basing his machines on second-hand tractors. Thomas, from Great Bourton near Banbury in Oxfordshire, was a dealer who exported used tractors. One of his largest markets was to the USA, where he was continually asked for tractors with more power. The idea to build six-cylinder conversions based on used Ford 5000 tractors is reputed to have come to him while on a flight from Seattle to Chicago in 1972.

Thomas approached Roadless who designed him a conversion kit, and supplied him with the necessary parts, including the bell housing, side-support members, axle mounting bracket and bonnet, to fit six-cylinder 2713E engines into Ford 5000 tractors. Apart from working for Roadless, Mervyn Ford was also a partner in Eagle Plant Hire at Warwick with his father and a third partner, Ralph Oliver, and undertook to build the tractors up for Thomas on the premises,

RIGHT: Thomas Ninety-Five 100.

BELOW: Thomas Ninety-Five 100 partly completed for cab testing at Silsoe.

working at nights in an 18 sq ft asbestos shed.

Thomas bought up as many used Ford 5000s with Select-O-Speed gearboxes as possible because, at the time, Ford were offering an exchange plan on these old epicyclic transmissions for new eight-speed manual gearboxes complete with a new crown-wheel and pinion. This gave a two-fold advantage to Thomas – the tractors with Select-O-Speed units were disliked and could be obtained cheaply, and his finished conversions ended up with new transmissions for no extra outlay. A number of Doe 130 tractors, consisting of two 5000 units in tandem, were also bought up and converted back to two tractors. All the tractors were reconditioned where necessary, and new 13 in. cerametallic clutches were fitted.

The first tractor was completed in April 1973, and sold through J. J. Thomas and Associates as the Ninety Five-100 with the benefit of 104 bhp from the 2713E unit. The company would supply complete tractors, or convert customers' own machines for them. The 114 bhp 2714E engine was also offered as an alternative power-unit. About thirty or forty machines were built at Eagle Plant Hire, which in 1978, through a change of name, became Axis Plant Hire, and is still run today by Mervyn Ford.

In the late 1970s, Thomas moved tractor production to Ironstone Works in Wroxton St Mary near Banbury, and formed J. J. Thomas (Farm Tractor Sales) Limited. In 1978, the company brought out a three-line air braking system, which was marketed as the Farmair kit at £750 for fitting to Ford tractors.

Thomas was now manufacturing complete new tractors using the Ford 7600 and 7700 skid-units. The Ninety Five-100 Mark ll was based on the 7600. Fitted with the 2714E engine and a straddle cab, it was priced at £12,547. Based on the 7700 and also using the 2714E engine, the Ninety Five-100 Mark lll had a flat-floor Clepa cab made in Holland, and cost £13,072. In 1979, a four-wheel drive version of the Mark lll was launched, fitted with a Schindler axle and priced at £16,327. The Ninety Five-100 Mark l continued as a conversion built on a second-hand unit for £9,850 complete, or the company offered to convert farmers' existing 5000, 7000, 6600 or 7600 tractors, and return them in about two weeks.

In about 1980, the Ninety Five-120 tractors came out, still using the 2714E engine, but now fitted with the same Lambourn flat-floor cab as used on the Roadless K Series. The company disappeared from view during the early 1980s, probably after competition from Ford made its products no longer viable, but not before it had produced well over 100 tractors, all using conversion kits supplied by Roadless.

TOP:
Thomas Ninety-Five 100 with a Clepa cab.

ABOVE:
Thomas Ninety-Five 120 with a Lambourn cab.

APPENDIX 3
Roadless Applications

I. Orolo Track Units (Locked Girder)

Denomination	Weight per track	Applications
DW1	2 cwt	Barrows, log carriers, portable engines, etc
DW3	6 cwt	As above. Heavier loads, agricultural machinery, etc
D1	10-15 cwt	Bullock carts, agricultural machinery
D2	1-1½ tons	Wagons, trailers, heavy equipment
D3	2-3 tons	Platform trailers, heavy equipment
D4	4-5 tons	Heavy equipment
D8	8-10 tons	Lifeboat carriages, cranes, heavy equipment

Note: G Series Orolo units with rubber-jointed tracks introduced 1931

Appendices

II. Flexible All-metal Tracks

Applications (Makes in brackets show actual uses)

B1
Colonial model half-track cars (Rover, Austin)
Military reconnaissance machines, staff cars (Wolseley)
All-track extra-light tractors, for hauling, ploughing etc
All-track carriers up to 2 tons

B2
Half-track lorries, 20-25 cwt (Guy, Morris Commercial)
Light armoured cars (Guy)
All-track light tractors to haul 4-5 tons
All-track carriers up to 3 tons

B3
Half-track lorries 2½-3 tons (AEC, Peugeot, Vulcan, Daimler)
Heavy armoured cars (Daimler)
Light tanks (Morris Martel)
Medium tractors to haul 8-10 tons, ploughing etc. (Peterbro)
All-track carriers up to 6 tons

B4
Half-track medium tractors (Sentinel, Fowler)
Half-track lorries to carry 4 tons (FWD)
Extra-heavy armoured cars
Medium tanks
All-track carriers up to 10 tons

C
Half-track steam wagons to carry 6-7 tons (Fowler)
Extra-heavy tractors to haul 20-25 tons
Heavy tanks
All-track carriers up to 15 tons

Notes: B-type tracks use rubber-tyred cast steel rollers mounted on leaf springs
C-type tracks use cable suspension system

III. E-type Rubber-jointed Tracks (Elastic Girder)

Tractor Applications

Tractor model	Make	Track type	Approx weight	Approx hp/ draw bar/ belt
Motor Cultivator MG 2	Ransomes	J	10 cwt	5/-
Bristol Tractor	Bristol	E	1 tons 1 cwt	12/10
Roadless FS	Fordson N	E3	2 tons 16 cwt	15/20
Roadless FE	Fordson N	E3	3 tons 0 cwt	15/20
Roadless FH	Fordson N	Heavy-duty	3 tons 0 cwt	15/20
Roadless CS	Case C	E3	3 tons 0 cwt	20/25
Roadless CE	Case C	E3	3 tons 4 cwt	20/25
Roadless RS	Rushton	E3	2 tons 18 cwt	16/26
Roadless RE	Rushton	E3	3 tons 2 cwt	16/26
Roadless MS (diesel)	Marshall	E3	3 tons 7 cwt	18/30
Roadless ME (diesel)	Marshall	E3	4 tons 10 cwt	18/30
Heavy-duty Roadless L	Case L	Heavy-duty	5 tons 0 cwt	35/40
Lifeboat Roadless	Case L	Heavy-duty	6 tons 0 cwt	35/40
Roadless G (diesel)	Garrett	E3	4 tons 15 cwt	42/60
Roadless MHE	Massey Harris 12/20	E3	3 ton 0 cwt	12/20
Roadless MHS	Massey Harris 12/20	E3		- 12/20
Heavy-duty Roadless MH	Massey Harris 25/40	Heavy-duty	4 tons 15 cwt	25/40
Roadless ACR (diesel)	Fordson/Ailsa Craig	E3	-	-
RT 50	Ransomes & Rapier	Heavy-duty	8 tons 0 cwt	65/-
MAVAG Roadless	MAVAG	E3	4 tons 10 cwt	-
Model E Roadless	Fordson E27N	E3B	-	30/-
Roadless J17	Fordson E1A	J	4 tons 10 cwt	52/33

Notes: prefix S for standard model
prefix E for extended model
prefix H for heavy-duty model
J-type rubber-jointed tracks used on Orolo track units, Fordson E1A J17, Roadless R20 and Ransomes MG Crawlers
– indicates no available information

IV. DG Half-tracks (Driven Girder)

Applications (Date of introduction in brackets)

DG4 (1943)
Fordson N, Fordson E27N Major, Case C, Case DEX, Marshall 12/20, Field Marshall Series II, David Brown Cropmaster, Landini L45, Steyr, Lanz Bulldog, Fordson E1A Major

DG4A (1946)
Oliver 80, John Deere D, Massey Harris 102 Junior, Massey Harris 744 D

DG5 (1943)
Fordson N on 52 in. front tyres

DG6 (1944)
McLaren Motor Windlass, FWD trucks

DG7 (1944)
Allis Chalmers Model U

DG8 (1944)
Case L, Case LA, Oliver 90

DG13 (1946)
Ferguson, Ford Ferguson 2N, Ford 8N, Ford NAA, Allis Chalmers B, Newman, Landini L25

DG15 (1953)
Fordson E1A Major, Nuffield Universal, Landini L55

Note: DG9 (rubber-jointed), DG10, DG11, DG12, DG14 & DG16 tracks also designed

V. Roadless Four-wheel Drive Tractor Production

Tractor model	Built from	To	Number built	Engine	Transmission	Comments
Major	March 1956	December 1964*	2947	E1A Major	E1A Major	All 4-cylinder production inc. Power Major and Super Major
Ploughmaster 6/4	March 1962	October 1964	200	590E	E1A Major	
Ploughmaster 6/2	May 1961	December 1964	23	590E	E1A Major	Two-wheel drive for export
Dexta	January 1960	November 1964*	78	Dexta (Perkins F3)	Dexta	Production includes Super Dexta Conversion supplied by Selene
International B450	October 1963	December 1970	379	I/H B450	I/H B450	Kits supplied to Doncaster
Ploughmaster 65	December 1964	August 1968	1000	5000 (6X Model)	5000 (6X Model)	
Ploughmaster 90	June 1965	June 1966	127	590E	5000 (6X Model)	
95	June 1966	December 1974	214	2703E	5000	2713E engine fitted 1971 onwards
Ploughmaster 46	February 1966	November 1972	45	3000	3000	Conversion supplied by Selene
Ploughmaster 80	May 1967	April 1968	24	5000	5000	CAV Turbocharger
Ploughmaster 75	September 1968	October 1975	1188	5000 (6Y Model)	5000 (6Y Model)	
115	February 1968	May 1975	182	2704E	5000	2714E engine fitted 1971 onwards
115 High Clearance	January 1969	July 1976	43	2704E	5000	High clearance for Puerto Rico
120	December 1971	December 1983	51	2715E	5000	2725 engine fitted 1982 onwards
94T	December 1971	November 1975	146	7000	7000	Turbocharged
International B-614	April 1968	October 1968	50	B-614	B-614	Kits supplied to Doncaster
International 634	June 1969	July 1971	218	634	634	Kits supplied to Doncaster
International 444	August 1973	March 1975	80	444	444	Sige axle Kits also supplied to Manitou
105	July 1974	August 1976	34	2714E	5000	
Ploughmaster 78	November 1975	June 1983	356	6600	6600	
98	December 1975	May 1982	150	7600	7600	Turbocharged
118	June 1976	September 1982	91	2715E	7600	
780	August 1979	December 1981	12	6600	6600	Portal axle
980	September 1979	December 1989	16	7600	7600	Portal axle Turbocharged

Notes: * Kits supplied after this date.
Date of first built is for production models. Prototypes may have been built before this date
Numbers built include conversions supplied as kits
The transmissions used in the tractors were often modified or strengthened by Roadless

Useful Addresses

**Agrosave, The Sidings, Abbots Salford Road,
Harvington, Evesham, Worcs, WR11 5XJ
Tel: 01386 870730. Fax: 01386 871133**
Genuine and original Roadless tractor parts, also Selene spares.
Trade and export.

**Four Wheel Traction,
15, North Audley Street, London, W1Y 2LR
Tel. 0171 629 9966**
Power steering kits for Ford tractors.
Reduction boxes for Massey Ferguson tractors.
Spares for Four Wheel Traction conversions.

**Robin Ketley, Bronwylfa Farm,
Mynydd Isa, Mold, Clwyd, CH7 6TF**
Owner's register for Selene Massey Ferguson conversions.

Bibliography

This book could not have been written without recourse to the many publications that were produced in the name of Roadless Traction, including sales brochures, advertising booklets and manuals of various types. There was never any shortage of printed material coming out of Hounslow, much of it bearing the imprint of Philip Johnson, and in later years, Ven Dodge. The company was nothing if not adept at documenting its own work. *Roadless News*, especially, is an important and valuable record of transportation and agricultural engineering developments, not just at Roadless, but across the world. It is also recommended as an enlightening read on many diverse subject matters other than technical innovation. I am also grateful for the reams of company correspondence which I have very kindly been allowed access to, yielding much important background information. All the following publications have also proved useful sources.

JOURNALS & OTHER PERIODICALS
AFV Series
Agricultural Machinery Journal
Arable Farming
Classic and Vintage Commercials
Farm
Farmers Weekly
Forestry and British Timber
Implement and Farm Machinery Review
Land Rover Owner International
Military Modelling
Power Farmer
Motor Transport
Vintage Commercial Vehicle

BOOKS & OTHER PUBLICATIONS
Baldwin, Nick, **Vintage Lorry Album,** Marshall Harris & Baldwin Ltd. 1979.
Baldwin, Nick, **Vintage Tractor Album,** Marshall Harris & Baldwin Ltd. 1980.
Baldwin, Nick, **Vintage Tractor Album Number Two,** Frederick Warne 1982.
Batchelor, John H., and Macksey, Kenneth, **Tank,** Macdonald & Jane's 1970.
Cornwell, E. L., **Commercial Road Vehicles,** Batsford 1960.
Fuller, Colonel J. F. C., **Economic Movement,** The Tank Corps Journal 1922.
Fuller, Colonel J. F. C., **Pegasus,** Kegan Paul, Trench, Trubner & Co. Ltd. 1925.
Hughes, W. J., A **Century of Traction Engines,** David & Charles 1970.
Hughes, W. J., and Thomas, Joseph L., **The Sentinel,** David & Charles 1973.
Olyslager Organisation, **Half-Tracks,** Frederick Warne 1971.
Sherwen, Theo, **The Bomford Story,** Bomford & Evershed Ltd. 1979.
Smithers, A. J., **A New Excalibur,** Grafton Books 1988.
Smithers, A. J., **Rude Mechanicals,** Grafton Books 1989.
Trythall, Anthony John, **Boney Fuller–The Intellectual General,** Cassell 1977.
Wheldon, John, **Machine Age Armies,** Abelard-Schuman, 1968.
Wittering, W. O., **The Hydrostatic Skidder,** Forestry Commission HMSO 1974.

Index

Roman numbers refer to text, bold to illustrations

A

AEC Y-Type 4-ton lorry, **27**
Agrosave, 170
Ailsa Craig Roadless (ACR) tractor, 45, **47**
Allis Chalmers
    Model 'B', 71-2, **72**
        orchard tractor, 72, **73**
    Model 'U', 68, **67**
Angier, David, 148, 154, 160
Austin 20 hp, **28**
Axis Plant Hire, 179
axles
    high clearance, 80-83, 125, **84, 125**
    Manuel Roadless, 96-7, 98, **92**
    planetary, 99-100, 129-30, 134, 141, **117-18, 134-5**
    portal, 145-8, **148-9, 166**
    Schindler, 145-6
    Selene, 176, **175, 177**
    'wide axle', 98-9

B

Barford & Perkins Roadless tractor, 41, **40**
Battelle, Arthur, 111, 145, **134**
Batty, Sir William, 92-3
Benson Group, 168
Booth, Cecil, 27-8
Booth, Rupert F.C., 110, 148, **110**
Bristol crawler, 53-6, **54-7**
Bristol Tractors Ltd, 54-6
Bruff TG3, 125-7, **130**

C

cable suspension system, 16, 18-19, 20-22, 25, 29, **10, 12, 15-18**

Case
    DEX, 63-5, 68, **63-4, 66**
    LA, **66**
    LH, 46, **50**
    Lifeboat tractor 47-8, **50-51, 58-9, 76-7**
    Model 'C', 46, 68, **48-9, 65**
    Model 'L', 46, 48, **48-9**
Clark, Lt-Col Charles Willoughby, 18, 20-22, 24
Cleveland Cletrac HG, **82**
Collins brothers, 163, 164
County
    '744', **162**
    equal-wheel tractors, 122
Crockford, Vic, 110-11, 148, 157, 168, **127, 155**
    designs, 123, 129, 131, 154, 157, 160
Crossley open tourer, 21, **14-15**

D

Daimler Roadless trucks, 33, **30-31**
Dare-Bryan, Valerian, 163
David Brown Cropmaster, **68**
Davis, Major Lewis K, 19, 25
Dodge, Alfred Ventham ('Ven'), 110, 154, 167, 168, **110**
Doe
    '130', 179
    '5100', 178
Duncan cabs, 129, 140, **117, 122, 131-2, 151, 165**

E

Eden Manuel conversion, 175, **174**

F

Farmair braking system, 179

Ferguson
    FE35, 175, **173**
    plough, **54**
Field Marshall
    Series II, 71, **72**
    Series III, **70**
Foden
    6-ton wagon, 29, **22-3**
    D-Type steam tractor, 29, **23**
Ford, Mervyn, 131-3, 142-4, 161, 168, 178-9, **134, 161, 164**
Ford
    6X range, 112-13, 176, **177**
    '7000', 176, **177**
    '8000', 130, **135**
Ford Ferguson
    2N, 72, **72**
    8N, 72
    NAA, 72
Fordson, 57
    Dexta, **97**
    E1A Major, 73-4, 83, 84, **75-6, 84, 86-7, 90, 93, 172**
    E27N Major, 69, 80-83, **68-71, 83-4, 90, 172**
    orchard tractor, **69**
    rowcrop conversion, 84-7, **85-7**
    Major, 100, **95**
    Model 'N', 69, **64-5**
    Power Major, **88-9, 94-5, 99**
Fordson Roadless
    full track, 43-5, **43-7, 58-9, 61**
    half track, 63-4, **58-9, 61-3**
    orchard tractor, 45, **45**
    swamp tractor, 45, **44, 46**

Index

Forestry Commission, 154-9, 161, 162
Four Wheel Traction Ltd, 175, 177
Fowler 'Snaketrac', 29-30, **24**
Fuller, Lt-Col John Frederick Charles, 15, 26-7
Fuller, William 172-3, 175, 177
FWD
    lifeboat tractor, 68, **32**
    trucks, 33, **31-2**

G

Garrett Roadless tractor, 49-51, **38-9, 52**
Gibbon Roadless lime spreader, 74-5
GMC 2½-ton 6x6 truck, **92**
Guy
    armoured car, **30**
    truck, 33, **29**

H

H. C. Slingsby, 37
Haynes, Roger, 144-5, 170

I

International
    '384', 131, **136**
    '444', 130, **136**
    '574', 131
    '634' Four-Wheel Drive, 127, 130, **131**
    '674', 131
    B-450 4-Wheel Drive, 103-4, 127, **104**
    B-614   127, **131**

J

J. J. Thomas and Associates, 179
J. J. Thomas (Farm Tractor Sales) Ltd, 179
Jewell Logtrac, 169, **169**
Jewelltrac
    '103', 169, **168**
    '116', 169
    '120', 169, **168**
John Allen & Sons (Oxford) Ltd, 37
John Fowler & Company, 12-13, 16-17, 20
Johnson, Lt-Col Philip Henry, **10, 65, 80, 90**
    character and early career, 10-13
    death, 110
    and Roadless Traction Ltd, 17-18, 23-4, 26-7
    and Selene SAS, 90-93
    Tank Corps developments, 13-15, 18-23

K

Kellands Group, 170

L

L. F. Jewell Ltd, 168, 170
    Roadless Division, 168-70
Lambourn cabs, 140, 141, 142, **142, 152, 179**
Landini
    L25, 72
    L45, 72, **74**
    L55, 72
Lanz Bulldog tractor, 72, **75**
Leyland four-wheel drive tractors, 175
Long Sutton tractor trials, 131-6, 142-4

M

Manitou fork lifts, 131
Manuel-Roadless Fordson Major, 96-7, 98
Marshall Roadless 18/30, 51, **52**
Massey Ferguson
    '35', 174
    '65', 175, **173-4**
    '135', 175, **174**
    '165', 175, **175-6**
    '175', 175, **175**
Martel, Major Giffard Le Q, 35
Massey Harris
    12/20, 48
    25/40, 48
    102 Junior Twin-Power, 68, **67**
    744D, 72, **73-4**
Massey Harris Roadless, **51**
Mavag, 52, **53**
McConnel-Thornton-Garnett TG3 trenchless drainer, **130**
Melvin, Graham, 168-9
Morris
    Commercial Roadless half track, 34-6, **8-9, 33-4**
    Martel tank, 35-6, **35-6**

N

Northrop tractor, 122
Nuffield Universal DM4, **76**

O

Oliver 80 Standard 68, **67**
Orolo track units, 28-9, 31, 60, **18, 21**

P

Peterbro Roadless, 40, **40**
Peugeot 4-ton truck, **28**

R

Rackham, Capt G. John, 17, 21, 25
Ransomes & Rapier RT50 crawler, 51-2, **53**
Ransomes
    MG crawler, 56-7
    MG5, **56**
Roadless
    6/2, 102-103, **165**
    25 toolbar, 57, **56**
    '78', **148**
    94T, 129, 134, 140, **133-4, 137**
    '95', 130, 134, 144-5, **134-5, 137, 144**
    '98', 140, 165, **140-41, 144, 146, 152**
    100-ton road train, 36-7, **37**
    '105', 130, 140, 141, **137**
    '115', 124-7, 129, 134, 140, 144-5, **108-9, 123-8, 130, 170**
        extended version, **130**
        high clearance, **108-9, 129**
    '118', 140, 142-4, 145-8, 165, 167
    '120', 129, 141, **133, 137, 142, 146, 167**
    '700' loading shovel, 120-21, **120**
    '780', 146-8, **148**
    '980', 146-8, **148-50**
    '9804', 165, **166**
    Amex Workhorse, 163, **163**
    bicycle trailer, 107, **107**
    carts, 28, **20**
    Dexta, 100, **97-8**
    four-wheel drive conversion, 96-106, 131, **88-98, 104, 131, 136**
    hydrostatic forestry tractor, **152**
    J series, 142, **143-6**
    J17 crawler, **78-9, 80-82**
    K series, 142, **146-7**
    Land Rover, 104-7, **99, 105-6**
    land torpedo, 61, **60**
    Lifeboat tractor, 47-8, **50, 58-9, 76-7**
    loading shovel, 99
    Logmaster, 154, 160-63, **138-9, 158-62**
    Mack truck, **18**
    Model 'E' crawler, 77-8, **77-80**
    Motorcycle, 19, **12**
    Ploughmaster 6/4, 101, **99-103**

Ploughmaster 6/4 Super 90, 101-2, **103**
Ploughmaster '46', 117, 121, 130, **118, 122**
Ploughmaster '65', 113-15, 117, 119-20, **111-13, 119-21**
Ploughmaster '75', 121, 130, 140, **121, 123, 128, 131-2**
    high clearance, 125, **129**
Ploughmaster '78', 140, 165 **140-43, 147, 151**
Ploughmaster '80', 117-19, **118**
Ploughmaster '90', 101, 114-17, **114-5**
Ploughmaster '95', 117, 121, 129, **116-17, 121-2, 132**
Power Major, **95-6**
R20 crawler, 79-80, **83**
S series, 165
Six-2, 165-6
skidders, 154-63, **153-7**
Stretcher, **18**
Super Dexta, **96, 98-9**
Super Major, 99, 100, **96, 98-9**
Teleshift TS230, 163-4, **164**
Teleshift TS430, 164, **164**
tricycle rowcrop conversion, 84-7, **85-7**
Type C barrow, 28-9, **19**
tyres, 36, **37**
Roadless Holdings Ltd, 152, 167
Roadless Patents Holding Company, 25
Roadless Traction Ltd (1919-1980), 17-18, 93-6, 152
    directors and personnel, 17-18, 23-4, 110-11
    Gunnersbury House, Hounslow, 25-6, 37, 60, 127, 148-9, **19**
    Hounslow Assembly Shop and Yard, **71, 99-100, 143, 151**
Roadless Traction Ltd (1980-1983), 152, 167, 168, 170
Robert Eden and Company, 172, 175
Rover Car company prototype tractor, 52-3, **54**
Rover Colonial, **27**
RTL Engineering, 152, 167
    Sawbridgeworth Works, 149-52, **150**
Rushton Roadless, 42-3, 62, **41-3, 60**
RWC (Developments) Ltd, 163, 164, 168

S

Samson Roadless, **17**
Samson Sieve Grip '12-25', 25, **16**

Sanderson Teleporter, 163
Searle, Col Frank, 14, 52
Segre-Amar, Dr Sion, 91, 92, 93, 172
Selene
    4RU, 176, **177**
    four-wheel drive conversions, 91-6, 172-7, **172-5, 177**
    MF175 4RU, **177**
    transfer box, 173
    Selene SAS, 91, 93-6, 172
Sentinel-Roadless tractor, 30-31, **24-5**
Servis-Roadless brushcutter, 83-4
Shaw, Lt, 17
Skelton, Charles William, 14, 21, 24, 110, **57, 65, 80**
Steyr tractor, 72, **75**
Super-Sentinel steam tractor, 30-31, **24-5**

T

tanks, 13-15, **11-16**
    Mark V, 15, 16
    Martel, 35-6, **35-6**
    Medium A Whippet, 15
    Medium D, 15, 16-17, 19-20, 22-3, **11, 13-14**
        D Star, 17, **11**
        D Two-Star, 17, **11**
        Light D, 18-19, **12**
        Light D Star 18-19, **12**
    DM, 22, **15**
Teleshift '404', 168
Thomas, J. J., 178-9
Thomas
    Ninety-Five '100', 179, **178-9**
    Ninety-Five '120', 179, **179**
Torchio, Signor, 175
tracks, 16, 18, 19, 24-5, **12, 16-17**
    B-type, 29
    C-type, 29
    Driven Girder (DC) half-tracks, 67-75, **65-76**
    Elastic Girder (E) rubber jointed, 41, 75-6, 78-9, **41**
    Orolo, 28-9, 31, 60, **18, 21**
    skeleton, 69-71, 74, **69, 72-4**
    'snake track', 22-3, 25, 29-30, 36, **21, 24**
Tripp, Leonard William, 24, 57, 110, 122
Trusty garden tractor, 68, 73

V

Vulcan 2-ton lorry, **28**

W

Walter Lawrence Engineering, **149-50**
'Waveless' road roller, 31, **26**
White Staff Observation Car, 19, **13**
Whitlock Dozaloda, **96**
Wolseley car, **29**

Other Books by Stuart Gibbard are:
Tractors at Work Volume 1
Tractors at Work Volume 2
Ford Tractor Conversions

FARMING PRESS BOOKS & VIDEOS

Below is a sample of the wide range of
agricultural and veterinary books and videos we publish.

For more information or for a free illustrated catalogue
of all our publications please contact:

Farming Press Books & Videos
Miller Freeman professional Ltd
Wharfedale Road, Ipswich IP1 4LG, United Kingdom
Telephone (01473) 241122 Fax (01473) 240501

Tractors at Work Volumes 1 and 2

Stuart Gibbard sets the tractor firmly in its working environment, showing in these two outstanding compilations of photographs as much about the recent social history of farming as about machinery development.

Ford Tractor Conversions: the story of County, Doe, Chaseside, Northrop, Muir-Hill, Matbro and Bray

Stuart Gibbard's detailed, profusely illustrated history of these companies and the models they produced based on the Ford tractor skid unit.

Ferguson Implements and Accessories

Written by John Farnworth, this is the first comprehensive book about the wide range of implements and accessories marketed under the Ferguson badge starting with the Ferguson Brown and finishing with the Massey-Harris-Ferguson. Over 370 illustrations of equipment are accompanied by basic technical details.

Books and Videos by Michael Williams

Farming Press have published five books by Michael Williams and four videos in which he is the presenter. Topics include Ford & Fordson, Massey-Ferguson and John Deere as well as the best selling *Tractors Since 1889* and the children's *Tractors: How They Work and What They Do*. His most recent video is *Henry Fords' Tractors 1907-56* which tells the story of Ford and Fordson from the 1907 prototype to the Fordson Major diesel, featuring tractors and their owners in the USA and Britain.

Classic Farm Machinery Volumes 1 and 2 and Classic Tractors

Compilations of archive video chosen and expounded by Brian Bell to show the development of agricultural machinery. Volume 1 covers 1940-70, Volume 2 1970-90 and *Classic Tractors* focuses on the development of tractors from 1945 to the present. '... you will never tire of watching ... I would not hesitate to recommend to anyone with an interest in farm machinery'
Farm and Horticultural Equipment Collector.

World Harvesters
Bill Huxley

Photographs and short descriptions of a wide range of harvesters of all types from all over the world.

Fifty Years of Farm Machinery
Fifty Years of Garden Machinery
Brian Bell

A pair of books which illustrate and describe the course of rural machinery development in Britain from the 1940s to the present. A host of models and manufacturers are dealt with, the emphasis being on the 1950s and 60s when progress was most rapid.

Fordson: the story of a tractor

This video narrated by Bob Symes, features the five main Fordson Models from 1917 to the 1950s. It combines archive material with new film.

Early and More Years on the Tractor Seat
Arthur Battelle

Two humorous autobiographical accounts of one man's life with machinery from a pre-war Fordson N and an IH W30 to the Fordson Dexta and Major.

Farming Press Books & Videos is a division of Miller Freeman Professional Ltd which provides a wide range of media services in agriculture and allied businesses. Among the magazines published by the group are *Arable Farming, Dairy Farmer, Farming News, Pig Farming* and *What's New in Farming*. For a specimen copy of any of these please contact the address above.

ROADLESS